Building a Lean Service Enterprise

Reflections of a Lean Management Practitioner

Building a Lean Service Enterprise

Reflections of a Lean Management Practitioner

Debashis Sarkar

CRC Press is an imprint of the
Taylor & Francis Group, an **informa** business
A PRODUCTIVITY PRESS BOOK

CRC Press
Taylor & Francis Group
6000 Broken Sound Parkway NW, Suite 300
Boca Raton, FL 33487-2742

Printed on acid-free paper
Version Date: 20160819

International Standard Book Number-13: 978-1-4987-7959-3 (Hardback)

Library of Congress Cataloging-in-Publication Data

Names: Sarkar, Debashis, author.
Title: Building a lean service enterprise : reflections of a lean management practitioner / Debashis Sarkar.
Description: Boca Raton, FL : CRC Press, 2017. | Includes index.
Identifiers: LCCN 2016025996 | ISBN 9781498779593 (hardback : alk. paper)
Subjects: LCSH: Service industries--Management. | Organizational effectiveness. | Cost control. | Quality control. | Industrial management.
Classification: LCC HD9980.65 .S268 2017 | DDC 658.4/013--dc23
LC record available at https://lccn.loc.gov/2016025996

Visit the Taylor & Francis Web site at
http://www.taylorandfrancis.com

and the CRC Press Web site at
http://www.crcpress.com

Printed and bound in the United States of America
by Edwards Brothers Malloy on sustainably sourced paper.

Contents

List of Figures

List of Tables

Preface

It has been an interesting journey since I embarked on a mission to embed Lean management practices in service companies. Over the past decade and a half, my effort has been to practice, propagate, and popularize Lean management as an effective approach for business excellence. It has been an enriching experience, and in the process I have learned and unlearned so many things that go into building a Lean service enterprise. I can tell you from my experience that a journey of Lean transformation can be challenging, sometimes even frustrating, but in the end it is rewarding. You have a transformation delivered by the people, for the people, and of the people.

When my books *5S for Service Organizations and Offices* (ASQ Press) and *Lean for Service Organizations and Offices* (ASQ Press) were published a decade ago, there were few service companies that talked about Lean. It is heartening to see that over the past decade things have moved in the right direction, and quite a few service companies have successfully used Lean tools and techniques for performance improvement. Experts and practitioners have shared their experiences through books and articles that have come in handy to those in the trenches. Some of the global consulting companies also have come up with approaches that have all added to the existing book of knowledge. And, one cannot forget the contribution of universities and schools that have dedicated researchers perusing this subject. All these have resulted in enhanced competency in this domain and made practitioners much more confident than they were a decade ago.

However, I still see many service Lean efforts falling by the wayside because of missing out on finer details. What we need today is many more experiences from the trenches to bring out those known-unknowns. I call them known-unknowns because these knowledge nuggets are known to a few that many others do not know but need to know. This book is an effort in this direction and endeavors to bring out some of those known-unknowns. It has ideas and lessons that will complement the existing cache of knowledge and can be utilized by both practitioners and those uninitiated. While practitioners will receive powerful firsthand experiences from a wide range of service contexts, executives looking at Lean as a potential approach for enhancing organizational effectiveness will get an idea of what it is like to be on this journey.

Although meant for service enterprises, the ideas are universal and can be adopted by manufacturing companies.

I look forward to your feedback. Do share with me your experiences, trials and tribulations, and challenges in building a Lean enterprise. My e-mail is debashis.sarkar@proliferator.net or debashis@debashissarkar.com.

Debashis Sarkar
www.debashissarkar.com

Notes to Readers

- *Lean professional* refers to Lean Change Leader, Lean Maven or Lean Expert, Lean Navigator, Lean Infrastructure Leader, and Lean Capability Leader.
- For uniformity, the book does not differentiate between a he and a she.
- The DEB-LOREX™ model is a trademark owned by the author.

The following tools have been created and successfully used by the author during Lean transformations:

- Deb's Questionnaire—For Taking Up a Lean Leader's Role
- Deb's 10 Commandments for Lean Professionals
- Deb's 15 Cs for Lean Transformation
- Deb's 10 Rules for Workplace Observations
- Deb's Lean Opportunity Questionnaire
- Deb's A-to-H Principles of Metrics
- Deb's Questionnaire to Determine Effectiveness of Process Optimization/Design
- Deb's Lean Health Assessment Tool
- Deb's Instrument for Assessing Visual Management Effectiveness
- Deb's 12 Wastes of Customer Acquisition

Author

Debashis Sarkar is one of the world's leading lights in Lean Management. Over the past two decades, he has been exploring and researching on how Lean manufacturing practices can be implemented in service organizations. He pioneered Asia's first Service Lean deployment in the early 2000s. He is credited to have proposed and deployed the world's first holistic blueprint for Lean for Service. He also designed and implemented the world's first 5S for workplace efficiency in an office setting. He has developed many new tools and techniques for Lean for Service, some of which appear in this book.

Debashis is the founder and managing partner of the boutique consulting company Proliferator Advisory & Consulting (http://www.proliferator.net), which enables companies with customer-centricity and Lean thinking.

Debashis has been invited all over the world for workshops and conferences and has authored 8 books and more than 70 articles/papers.

Debashis is a Fellow of the American Society for Quality and is a recipient of the Phil Crosby Medal in 2014.

Prior to getting into consulting, he held leadership positions in companies such as ICICI Bank, Standard Chartered Bank, Unilever, and Coca-Cola.

You can reach him at debashis.sarkar@proliferator.net | debashis@debashissarkar.com

You can also follow him on Twitter: @DebashisSarkar.

1

How Engaged Is Your CEO and Top Management?

So, you have been appointed to catalyze a Lean journey? But, do you know why there is this sudden interest? Is it because this is supposed to be the current fad or because there is a genuine need? Have you spoken to the chief executive officer (CEO) and others in the senior management team to get a pulse of their mind? It is well known that it would be futile to embark on a Lean transformation journey without the required sponsorship of management. I have seen organizations struggle despite spending tons of money on training and certification. This is because of lack of commitment from top management. I have come across leaders who will make tall claims about Lean in public platforms of power, but when it comes to spending time to understand Lean projects, they make themselves invisible.

A CEO hired a Lean expert to embed Lean thinking in a shared service center. (Shared service centers are entities wherein similar services or processes that were done in various parts of the organization are hubbed for deriving efficiencies from economies of scale, enhancing quality, and improving scalability.) The reason he did this was not because he believed in its power, but because it was something nice to have. This is something he picked from a shared service forum in which members from other companies had claimed that Lean is a must for making processes efficient and effective. The expert was more of a trophy hire for him, which he used to showcase to his bosses from the corporate office. This was to show how forward-thinking and committed he was to process improvement. For him, this was another of the things he thought he needed to do before he was elevated to another role in the group. However, when it came to really understanding what Lean could do and how he could really leverage it for better efficiencies, he was quite indifferent. The Lean expert tried hard to

TABLE 1.1

CEO Engagement on Lean Efforts

Engagement Levels	What Is It?
1 Committed	Looks at Lean as a business strategy, and the CEO drives it personally. Also spends time to understand the basics and what the levers are for successful deployment. The CEO attends all relevant review meetings on Lean efforts and not only mandates all but also tries to create suitable context for Lean Change Agents to embed holistic Lean thinking across the company.
2 Passive	Does not personally drive Lean efforts but is not averse to them. Knows the power of Lean but wants the Lean Change Agent to engage his or her direct reports and other parts of the firm. Wants employees to adopt Lean based on successful pilots and not because the CEO mandates it.
3 Indifferent	Is aware of Lean deployment in the company but does not talk or spend time on it as he or she does not see it as an enabler for business outcomes.
4 Talker	Talks tall about Lean in public platforms and employee forums but makes himself or herself invisible when the question of involvement is concerned.
5 Detractor	Believes approaches like Lean are a waste of time and do not add value to business, even voicing concerns on why it should not be used.
6 Helpless	Knows the power of Lean but is not able to push because of organizational politics and silo mentality among senior leaders.
7 Parochial	Understands the power of Lean but is not able to spend time on it as there are other pressing issues. Does not see Lean as something "important" for him or her or believes it is not the time for Lean adoption.
8 Cost cutter	Sees Lean as a tool for cost cutting. Wants cost cutting in the company but garbs it as "Lean thinking."
9 Skeptic	Doubts the power of Lean because he or she has seen it not working in a previous organization or workplace. Needs to be convinced on what Lean can do.
10 Trophist	Looks at Lean experts as trophies for showing to one and all. Does not leverage this competency despite having the requisite talent in the company.

engage him, but this leader did not bother. The Lean expert tried hard to engage other senior leaders in the shared service center, but all resisted any form of change. Meetings were held to review some of the Lean projects initiated by the Lean expert, but they were more of a ritual. The Lean expert raised critical issues concerning inefficiencies, but the CEO kept quiet. All the other leaders joined together against him as they did not want anyone to bring out inefficiencies that plagued the processes. This charade of Lean implementation went on until the Lean expert left the organization frustrated. He gradually took all the team members he had hired to implement the Lean agenda. By then, the CEO also had another job outside the company, and he moved on. The sufferer in this whole thing was the organization.

A Lean Change Agent should always ascertain the engagement levels of the CEO and senior management. I have seen various levels of CEO engagement in Lean efforts. Table 1.1 summarizes these.

I would suggest before you decide to take a role in embarking on a Lean transformation that you obtain a sense of the engagement levels of your CEO and top management levels. You will have to put in place required strategy to take them on board.

2

Spend the First 90 Days to Understand the Needs of the Company

Just because you are made responsible for Lean deployment does not mean you immediately begin to act. Without really understanding what the company needs, I have seen Lean Change Agents holding training sessions and start working on a few projects. This is not the right strategy to follow. My view is to take up the role for Lean deployment and to step back and understand what the company needs under its current context. Remember, Lean cannot be implemented for the sake of Lean. It has to positively impact organizational performance. I would suggest spending at least the first 90 days walking around the company and understanding what the key issues faced by the company are. There could be issues that are quite obvious, and there could be those that are not so explicit. Walk around the length and breadth of the company, talk to all and sundry.

As a matter of fact, before you take up the role, you should talk to the chief executive officer (CEO) or someone in top management to give you the liberty of ABWA: ascertaining by walking around. Talk to all-level employees, from those reporting to the CEO and those in the front lines and shop floors. Chat with outside stakeholders, such as nonexecutive directors, customers, vendors, analysts, and regulators (if possible). Go through the annual reports and type of claims that have been made. Read what the outside world says about the company. If possible, reach out to one or two of the biggest critics of the firm. Also, reach out to one or two of those who have good things to say about the company. Spend time with customers to understand what the current issues and what their expectations are. With the finance folks, understand issues concerning profit and loss. What does the chief financial officer (CFO) or the CFO's team members think should be done to better the performance of the firm? Get a sense from the shop floor workers and front liners on what the key things are that need to be done to improve organizational performance.

Having completed the 90 days of ABWA, I would suggest you list the top three Focus Areas that you believe need to be addressed wherein Lean can be of help. Make a presentation of your findings to the top management, which includes the CEO and direct reports. Obtain their views if your understanding of the issues aligns with what they think are important. See if the issues that you have raised align with the Strategic Objectives of the company. If they do not, it could mean any of the following:

- The Focus Areas you have chosen are wrong and hence do not align with the Strategic Objectives.
- The Strategic Objectives are wrong and do not align with reality.
- The Focus Areas are correct and the current Strategic Objectives are also correct; all that needs to be done is to make sure the Focus Areas are a part of the Strategic Objectives.

The first 90 days encompass one of the finest learning times for a Lean Change Agent. The agent gets to understand the organization and this helps create the subsequent agenda of Lean deployment. From my experience, an expert Lean Change Agent will always bring out issues that were not known to the company. This agent gathers from observations and conversations. At times, this can surprise the CEO and top management as they tend to think they do not know what is happening in the firm. Little do they realize the value that a Lean expert brings from his or her experience.

When you highlight the Focus Areas, do not list them around functions such as people, process, technology, vendors, risk management, finance, and so on. This can make the respective functional owner look small in front of others, which no one would like. This can also pit one senior leader against another as there are always political overtones in a senior management team, and there are issues that a Lean Change Agent as a newcomer will never get to know.

Instead, talk in terms of the "end-to-end process," which includes not one but multiple functional owners. Figure 2.1 shows a typical Focus Area that was presented by a Lean Change Agent after the agent's ABWA. The list was whole-heartedly accepted by the top management team, and the first phase of Lean deployment saw multiple functional owners sponsoring the execution. As you can see, multiple functional owners agreed to

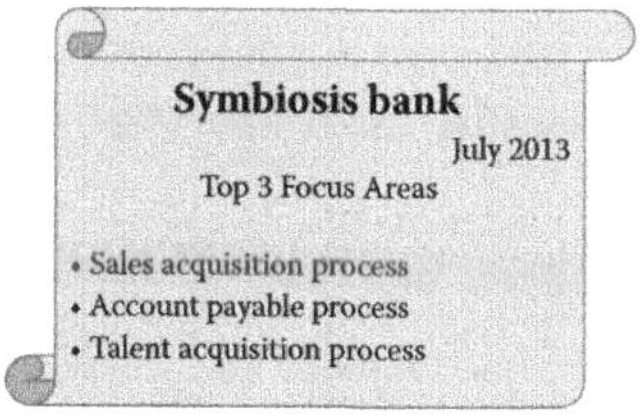

Symbiosis bank

July 2013

Top 3 Focus Areas

- Sales acquisition process
- Account payable process
- Talent acquisition process

Focus Areas	Business unit	Stakeholders*
Sales acquisition process	Mortgage	Head of retail banking, head of operations, head of credit
Account payable process	Entire bank (pilot in business banking)	CFO, head of operations, head of risk, head of business banking
Talent acquisition process	Entire bank (pilot in global markets)	Chief people office, head of global markets, head of talent acquisition

**Includes those who report to the CEO and will be attending your first meeting after 90 days of ABWA.*

FIGURE 2.1
Focus Areas in Symbiosis bank after 90 days of ABWA.

sponsor the first phase of deployments. In areas where the entire bank was affected, it was decided to do pilot tests in specific business areas.

These Focus Areas are critical as they are going to set the tone for Lean deployment later. My personal view is you should pick up the areas that have the maximum impact within a short period of time.

3

Should You Board the Ship?

Transforming an organization through Lean was Tim Noble's passion. He had been a successful change agent and had made a reasonable name for himself. He was working for an organization that did not seem to have used his talent to the fullest. He quit his last job out of desperation although he was paid well—his skills were not used. He had joined this organization with a lot of hope, but things did not turn out the way he wanted. He was told that he would be leading a major Lean transformation in a mortgage business and working toward embedding both efficiency and effectiveness, which could help to differentiate the organization in the marketplace. All these were promises made by the chief people officer (person responsible for human resources in a company) and the chief executive officer (CEO). Tim was doing well in a well-known manufacturing company, but the compensation levels and the pitch made by the CEO made him leave his job and join the mortgage company. He had spent nearly 10 years in the manufacturing company, and his Lean efforts there paid dividends. The vice president of manufacturing was committed to Lean, and Tim had all the support needed to embed Lean thinking across the organization.

He spent 2 years in the mortgage business but hardly managed any traction. The same was true for the Lean journey that was promised to start; he could not do even one meaningful project. All that Tim could do was map processes in the organization and some small local Lean process improvements. Since his was a high-cost hire, there were repeated murmurs in the company concerning the type of value that he was adding and if it was commensurate with the money he was being paid. In all his performance reviews, he was told that he was not adding value. Finally, after 2 years, he was told to leave the company as they did not see value in his work.

What happened to Tim Noble is representative of what happens when Lean Change Agents make the wrong decision in choosing which organization to join. In this case, there definitely were issues not only concerning inability to influence the organization but also concerning the challenge faced when a leadership "talks" about but is indifferent toward deployment. Also, challenges in service companies can be quite different from those in manufacturing companies. This is because "Lean" as a concept is still alien to many service companies despite "Lean for service" being on the horizon for more than a decade. Tim Noble would have fared much better had he stayed in the manufacturing company and not jumped ship. I have seen Lean Change Agents making wrong choices like Tim that they later regretted.

So, what should a Lean Change Agent do when making a decision whether to join a new company? The change agent should ask the 11 questions in Table 3.1; the responses in Table 3.2 should help the agent decide whether to sign up for a role or not.

In the mortgage company Tim had joined, if he had administered these questions, he would have received the responses shown in Table 3.2.

Tim Noble was a great tools person, but one thing he was not very good at was influencing skill. He was successful in the previous manufacturing company as he worked for a strong leader who was committed to Lean thinking. At the mortgage company, he was suddenly exposed to a set of

TABLE 3.1

Questionnaire: Taking Up a Lean Leader's Role

1. Why is the organization adopting Lean thinking?
2. What is the CEO's mindshare in the Lean effort?
3. Is this a greenfield deployment (new roll out of Lean) or is a scale-up of implementation under way?
4. What is the CEO's vision concerning Lean?
5. What are the top three deliverables expected in the first, second, and third years?
6. To whom will the Lean leader report? (If the Lean leader reports to the CEO, one knows the CEO is personally driving the agenda.)
7. What are the big change programs that are currently under way in the company?
8. Is the organization fine with a 90-day period of ascertaining by walking around (ABWA)? (One has to be a bit apprehensive if this is not allowed by the firm.)
9. What are the current challenges and key issues faced by the company?
10. What are the external factors and contextual changes that are affecting the organization? (For instance, if a merger is in the works, one has to do due diligence to find if there is an immediate appetite for Lean efforts and what the state will be once the merger happens.)
11. What would be the level of influence required to get the leaders on board?

TABLE 3.2

Questionnaire Responses

1. *Why is the organization adopting Lean thinking?* Its biggest competitor has adopted Lean thinking over the last 5 years and reaped huge benefits. There has been a lot of media coverage concerning it.
2. *What is the CEO's mindshare in the Lean effort?* This is difficult to ascertain, although in the interview the CEO talked about differentiating in the marketplace.
3. *Is this a greenfield deployment or is a scale-up of implementation under way?* Yes, this is a greenfield opportunity.
4. *What is the CEO's vision concerning Lean?* The company wants to differentiate itself in the marketplace by being more efficient and effective.
5. *What are the top three deliverables expected in the first, second, and third years?* This is something the Lean leader will have to work out once he or she joins the company.
6. *To whom will the Lean leader report?* The Lean leader reports to the head of technology and operations.
7. *What are the big change programs that are currently under way in the company?* The company is undergoing a major digitization to wean customers from branches to Internet and mobile.
8. *Is the organization fine with a 90-day period of ascertaining by walking around (ABWA)?* This is something the management does not support. They believe it is a waste of time. All senior hires are expected to be productive from the first day.
9. *What are the current challenges and key issues faced by the company?* The biggest challenge being faced by the company is failure to keep up with technology. It has been late in adopting technology compared to its competitor.
10. *What are the external factors and contextual changes that are affecting the organization?* Competitors have been eating into the organization's market share. It was a market leader 5 years ago.
11. *What would be the level of influence required to get the leaders on board?* The Lean Change Leader would need a large amount of influence to bring all the leaders on board.

leaders who did not know what Lean was all about and who had to be taken on board. The job was not that of a tool expert but of someone who could influence. Tim Noble could not ascertain the type of challenge that was in his new job. If he had a sense of the 11 questions, things could have been different.

4

Before Embarking on a Lean Effort, Pause to Understand the Type of Problem That You Are Trying to Solve

Because processes are an integral part of Lean transformation, Lean leaders have this knack of quickly playing with processes and working toward mapping and looking for inefficiencies. However, before you touch a process, I would recommend you step back and understand the type of problem that you are trying to solve. This will give you an idea of the issue at hand. Otherwise, you will put in a lot of effort but later realize it has not delivered what it was supposed to do.

So, what are the typical types of Lean efforts that one could pursue? Figure 4.1 summarizes these efforts.

Understanding the type of Lean efforts is important to make sure that not only are these efforts done right but also the right stakeholders are informed. So, for example, if a Lean leader's mandate is just sales force effectiveness of an auto loan process, the leader needs to focus on the end-to-end auto loan sales process and involve leaders from sales, operations, technology, credit, legal, and compliance departments. If the mandate is to work on holistic organizational transformation, the Lean leader needs to work on vision, mission, values, processes, behavior, technology, and so on. If the focus is on structural transformation, the focus will not be on process optimization but on role definition and activity rationalization. Understanding the type of Lean effort helps with clarifying the following:

- Scope of the problem
- Stakeholders to be involved
- Leaders who need to be influenced
- Duration

Type of Lean effort	What is it	Examples...
Local improvements	This pertains to local optimization work done to improve efficiency and effectiveness. This could be a part of a large process or just one of many locations.	• Removing waste in processes in a retail bank branch • Improving credit operations
Process transformation	This pertains to end-to-end design or optimization of a core process of an organization.	• Improvement of record to report process • Improving sales force effectiveness
Structural transformation	This pertains to fundamentally realigning the way the organization is structured for enhanced efficiency and effectiveness.	• Organizational simplification of food services business • Creating a process-based/value stream–based enterprise
Cultural transformation	This pertains to fundamentally changing the culture of the organization so that it's change ready for the future. The focus here is mainly leadership and people.	• Embedding customer-centric mindset among all employees • Building a culture of Lean thinking
Organizational transformation	This pertains to developing a new vision, mission, and values and radically changing the internal and external operations by focusing on all elements such as leadership, people, process, structure, etc.	• Holistic transformation of a mortgage business for better competitiveness

FIGURE 4.1
Types of Lean transformation.

- Type of capabilities that would be required
- Understanding the likely challenges
- Who will be affected by the change
- What emotions will have to be addressed
- What governance framework would be needed to monitor progress

5

Just Knowing Tools Does Not Make You a Lean Change Leader

The role of a Lean Change Agent is not for everyone. A person can know all the tools from the Lean toolbox yet may not be ready to take on the role of a Lean Change Leader. Knowing Lean tools can help a person obtain a certification, but if the person does not have executive presence and is not able to influence those around, the person will fail miserably. This was probably the reason why Tim Noble failed (Chapter 3) in a mortgage company. He did not have the presence and charisma that could influence a set of bankers. This is a big problem today. I have seen individuals who attend courses, do a couple of Lean projects, and claim they have become a Lean Change Leader. What makes a Lean Change Leader is not knowledge of tools but leadership depth. When a Lean Change Leader walks into a meeting room, one can feel his presence. (For ease of reading, gender neutrality is not maintained strictly throughout this work; however, both men and women can be Lean Change Leaders.) The Lean Change Leader guides organizations in their Lean journey and prevents them going astray. The leader's whole objective is to make Lean deployment self-sustaining.

So, what makes a great Lean Change Leader? Figure 5.1 summarizes the traits that make an effective Lean Change Leader. Let us understand what each of the five traits mean:

1. **Technical Skills:** A Lean Change Leader should have good knowledge of all relevant tools and techniques of Lean management. The leader should be someone who has used these tools during transformations and should have experience in at least three of the four types of Lean transformations discussed in Chapter 4. The Lean Change Leader should also have good project management skills and

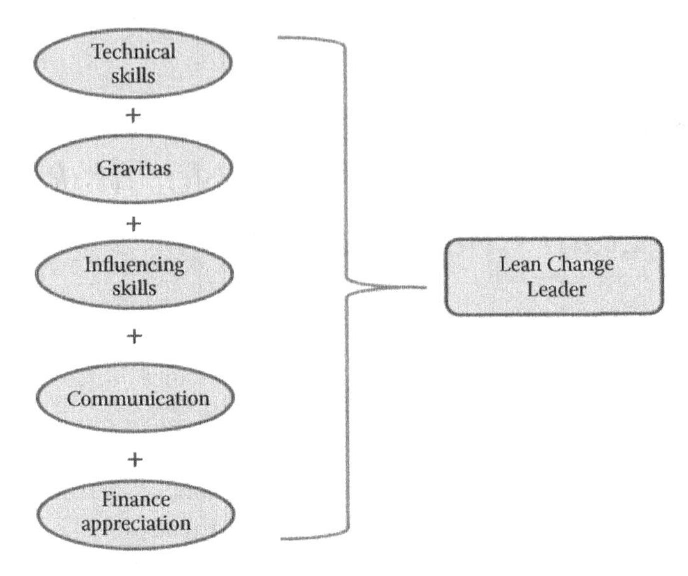

FIGURE 5.1
Traits of a Lean Change Leader.

a successful track record to coach a portfolio of projects. The goal of a Lean Change Leader is not expertise on the tools but how these tools can be used for business improvement under various contexts. If you want to know which tools a Lean Change Leader should know, I suggest you follow the recommendations prescribed by the Society of Manufacturing Engineers in partnership with Shingo Institute, American Society for Quality, and Association for Manufacturing Excellence.

2. **Gravitas:** Gravitas is a key trait a Lean Change Leader should possess. The leader should have "weight" and gravity so that all take him seriously. The Lean Change Leader has immense self-belief and belief in what he can deliver. Whatever the situation, the individual demonstrates immense confidence and is never rattled by ambiguity. There would be situations when a transformation may be complex, but the leader is able to quickly size up the issue and break it into smaller, manageable problems. Even when challenged by powerful leaders for the recommendations, the leader is able to stand his ground with composure. What drives the leader is the potential benefit that the organization can reap from these efforts. Whatever future state that the leader has envisioned, he is able to rally people around him and drive with a missionary zeal. The Lean Change

Leader never buckles under pressure and does not hesitate to show teeth for the larger cause of the enterprise. When it is necessary to choose between "maintaining good relations" versus "confronting on an issue," the leader chooses the latter for the larger benefit of the firm, knowing a Lean Change Leader's job is not about being popular but about doing the right things for the company. This leader is nimble, decisive, and action driven. His integrity is unquestionable, and his viewpoints emanate from deep-rooted, unshakable values. He is tough with those who block change yet sensitive in helping those who are struggling to manage change. At times, the leader may have to propose things that are unpopular, but he does it with humility. He is able to pick up signs of unarticulated emotions and knows when to be authoritative and when to be comforting. A Lean Change Leader can never be successful if he does not understand people's feelings as he knows sustained change happens through emotional engagement. An experienced Lean Change Leader would have a reputation that will precede him, but he knows every transformation is unique—past success does not guarantee future success. For the leader, every transformation is unique (even if they look similar), and the leader becomes involved with child-like curiosity and full dedication. A Lean Change Leader is a great listener. He spends a lot of time observing. He is known for thought-provoking questions and does not provide ready-made solutions. A Lean Change Leader is not born but is someone who has intentionally cultivated values and looks for opportunities to translate his convictions for organizational benefit.

3. **Influence:** This is an area that many individuals involved in Lean efforts do not get right. Of a seasoned Lean Change Leader's time, 75% goes into influencing others to adopt Lean practices for organizational benefit. This becomes further relevant when you are an in-house change agent trying to improve things. Gravitas and technical skills are required to influence others, and there is a bit of overlap. Some of us would say that "influence" is an integral part of gravitas, but I like to list it separately given its importance. A Lean Change Leader knows he has to influence those who do not report to him. Also, he has to work toward forming a new set of behaviors that help the progress of the Lean journey. Influence emanates from credibility, which comprises capability and character. Influence is discussed further in Chapter 7.

4. **Communication:** A Lean Change Agent needs to be an effective communicator. He communicates clearly, and when he opens his mouth, one can sense his clarity of thought. The Lean Change Agent comes across as someone who means business. Every word that he speaks is carefully chosen to create a positive impact on all. He is a good listener, and when he is with someone, he makes the person feel comfortable and heard. When the change agent speaks, he speaks with authority, and people listen. When addressing a group, he is quickly able to establish a connection with the audience. When speaking on specific issues plaguing the organization, he never shies from calling a spade a spade even if it means leaving a few high-and-mighty individuals red faced. If he has found potential problems, he brings them to the fore and has the courage to confront the relevant leaders. He never hides in verbosity but makes his communication direct and succinct. This is to make sure nobody is left guessing or misinterpreting what he said. Another trait that one sees in a Lean Change Leader when he communicates with a group is the ability to make his communication replete with stories and experiences so that all find it interesting and relevant.
5. **Finance Appreciation:** While one does not expect a Lean Change Leader to be a qualified accountant, he needs to have good understanding of various financial metrics that have an impact on business and how his efforts would affect the performance of the firm. He should have a broad understanding of key financial metrics and how they affect business. For example, he should know what a quick ratio is and that it measures a company's short-term liquidity. Another key thing that a Lean Change Leader should know is how processes link to the line items in a profit-and-loss statement. When a Lean Change Leader understands basic finance, he can quickly size up the key metrics of the business embarking on a transformation.

If you want to be an effective Lean Change Leader, focus not only on the toolbox but also on gravitas, influence, communication, and financial acumen.

6

Know the Building Blocks

Knowing the building blocks that make the organization is critical before one embarks on a Lean transformation. As a part of the 90-day ABWA (ascertaining by walking around), the Lean Change Agent should focus on the following: building blocks and key facets.

BUILDING BLOCKS

Building blocks are the functional blocks of the organization and give a view of how they are organized for delivery. Figure 6.1 is an example of the functional blocks of a finance organization captured during an ABWA.

KEY FACETS

Key facets of the business are the particulars of a business that the Lean Change Leader needs to obtain a view to realize a sense of the organization, its vision, its strategic priorities, and other elements that work together to deliver what the customer wants. To discover more about the key facets, refer to Figure 6.2.

With information on the building blocks and key facets, a Lean Change Agent can be more confident of the Lean transformation on which he plans to embark. Without this information, sometimes Lean practitioners get it wrong. For example, a Lean practitioner working on a retail banking transformation recommended removal of a process that later was found imperative for meeting regulatory requirements.

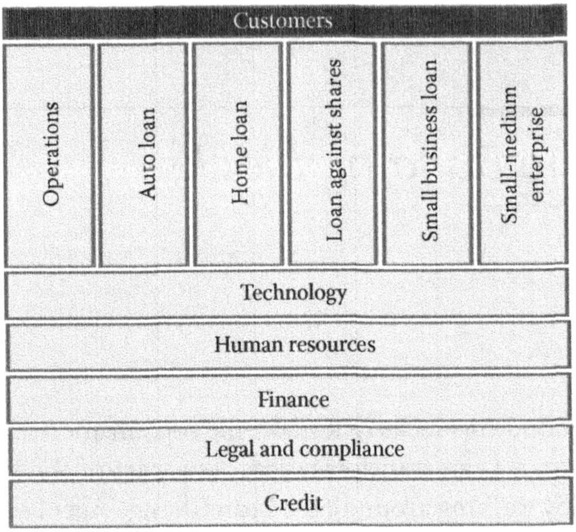

Customers

Operations | Auto loan | Home loan | Loan against shares | Small business loan | Small-medium enterprise

Technology

Human resources

Finance

Legal and compliance

Credit

FIGURE 6.1
Functional components of a finance company.

Vision	Where does the organization see itself in next 5–10 years?
Values	What are the values that drive the organization?
Customers	Who are the customers of the business and which customer segment do they cater to?
Products/ service	What are the key products and services of the business?
Strategic priorities	What are strategic priorities of the company for the next 12–18 months?
Structure	How is the organization structured for product and service delivery?
Core processes	What are the core processes of the organization?
Metrics	What are key strategic and operational metrics to ascertain organizational performance?
Incentives	What is the incentive framework to reward performance?
Regulations	What are the regulations that govern this business?

FIGURE 6.2
Key facets of business.

7

"Influence": Least Discussed Yet the Most Important Quality of a Seasoned Lean Change Leader

What is influence? According to one edition of the *Oxford Dictionary*, *influence* is "the capacity to have an effect on the character, development, or behavior of someone or something, or the effect itself." What is influence for a Lean Change Leader? For a Lean Change Leader, influence is the ability to engage one and all and shape their actions and behavior to bring about positive change that affects the efficiency and effectiveness of an enterprise.

This is easier said than done. I have seen many leaders not getting it right not because they do not want to do it but because they do not know what lever to press to get it right. It is so important to sharpen this trait to get this positive change right.

My experience as a Lean Change Leader has been influenced greatly by the work of leadership expert Linda Hill of Harvard Business School. What I present here is what I learned from her works and successfully adopted as a Lean Change Leader.

A Lean Change Leader has an uphill task. He has influence on those who do not report to him. He has to learn to get on board leaders at all engagement levels (see Chapter 2). It becomes a further challenge when there is passivity or no commitment from the chief executive officer (CEO) regarding Lean transformation. Here, the Lean Change Leader first has to take the CEO on board and then other members of the top management team before embarking on Lean efforts. It's easy to say that the CEO is not interested to what one wants to do. You cannot throw in the towel at this point. Actually, influencing all involved is the first test of the leadership depth of a Lean Change Agent.

Linda A. Hill, in her book *Being the Boss* (Cambridge, MA: HBS Press, 2011), said influence emanates from trust, which comprises competence and character. Adapting this thought in the context of a Lean Change Agent, I believe influence emanates from the agent's credibility, which is built on pillars of capability and character. We did cover the importance of technical skills in Chapter 5. But, capability is much more than plain "technical skills." Let us look at both of these pillars.

CAPABILITY

Capability is the ability to successfully deliver a Lean transformation. It is about knowing what to do, how to do it, and doing it while managing the organizational terrain and taking the relevant people on board. By taking people on board, the leader finds who is interested in the agenda and who are resisters. For a Lean Change Leader, the three elements of capability are technical knowledge, execution ability, and political competence:

- **Technical Knowledge:** This includes understanding Lean tools and techniques and having experience in what tool to apply where. While it is good to have expertise in key Lean tools, a Lean Change Leader is not expected to be adept in all Lean tools and techniques. However, for the tools for which the leader does not have expertise, he should know where to reach out for this expertise.
- **Execution Ability:** This is the ability to use the tools and techniques during Lean transformation. There are three dimensions here: the ability to use a wide range of tools across various types of transformation and across industry. A seasoned leader could have high execution experience across types of transformation, across types of industry, or across both. One may not find too many people who have exposure to both. Table 7.1 provides a tool for ascertaining a Lean Change Leader's execution ability. A seasoned Lean Change Leader's execution ability is explained in Table 7.2.
- **Political Competence:** Many of us look at political competence as a wrong thing to have. A few Lean practitioners have told me that they are just masters of the Lean toolbox, but when it comes to having political competence, they say it is not their cup of tea. Unfortunately, they do not realize that political competence is an

TABLE 7.1

Lean Change Agents' Execution Capability

Depth of experience	Rating
Experience in Lean tool usage	H
Experience in various types of transformation**	H
Exposure across industry*	L or M

*Industry includes sectors such as health care, banking, insurance, food services, industrial goods, consumer goods, etc.

**For types of transformation, refer to Figure 4.1 in Chapter 4.

Rate each of the above as high/medium/low

integral part of the Lean Change Leader capabilities. It is necessary to get the agenda going in the organization. So, what is political competence? Professor Samuel B. Bacharach of Cornell University, in his book *Get Them on Your Side*, defined political competence as follows: "Political competence is the ability to understand what you can and cannot control, when to take action, who is going to resist your agenda, and who you need on your side to push your agenda forward. Political competence is about knowing how to map the political terrain, get others on your side, and lead coalitions" (Avon, MA: Platinum Press, 2006, p. 6). For a Lean Change Leader, this would mean the ability to quickly scan the organizational terrain and understand who is with you, who is against you, what you can control, and what you cannot control in your Lean transformation effort. Then, it is working with individuals to remove the concerns, convince them of the worth of the effort, and take them on board so that they support the Lean journey. The focus here is to understand the specific resistors for the change. This includes those who make their concerns vocal and visible and those who are quiet and yet do not support your effort. From my experience, opposition to Lean efforts comes in various forms; these are summarized in Table 7.3. This can be quite handy in understanding opposition to a Lean Change Leader's agenda.

TABLE 7.2

A Seasoned Lean Change Leader's Execution Capability Matrix

Depth of experience	Rating	Depth of experience	Rating	Depth of experience	Rating
Experience in Lean tool usage	H	Experience in Lean tool usage	H	Experience in Lean tool usage	H
Experience in various types of transformation	H	Experience in various types of transformation	L or M	Experience in various types of transformation	H
Exposure across industry	L or M	Exposure across industry	H	Exposure across industry	H

TABLE 7.3

Reason for Resistance to Lean Efforts

Specific concerns	Yes/No
Objectives and outcomes that Lean transformation is trying to achieve	
Lean as an approach for business improvement or its ability to solve the problem in hand	
Leaders don't like to call the improvement effort as Lean	
Intention and capability of team put together for the transformation	
Lean change leader's ability to deliver	
Dislike for Lean change leader due to past experience	
Leaders impacted by Lean efforts believe they will be shown in bad light	
Relationship and political issues among leaders	
There are other competing priorities	
Others	

CHARACTER

Character refers to the belief of the Lean Change Leader that what he does is always in the interest of the organization. He treats all people with respect and never recommends what he does not practice or follow. He cares for people, which does not mean he is "soft." People know he practices tough love to get things done for the organization. At times, he does have to recommend tough decisions, but he does it with grace and humility. His ego never comes in the way of organizational improvement, and he is open to feedback, to change, and even to saying "sorry" if he is wrong. Do not forget that the Lean Change Leader needs to have values such as integrity, ability to collaborate, and passion for the organization and its customers; these are nonnegotiable.

8

Engagement: Where to Begin?

Having been appointed as a Lean Change Agent, a person often struggles with which approach to follow to get the initial engagement going so that Lean practices are adopted for business improvement. Although we did discuss the pillars of "influence" in Chapter 7, we need to know what one should do in the early days of a Lean journey.

There is no "one-size-fits-all" strategy to get this done. I have used a mix of approaches that have worked in various settings. However, what is important is to understand the audience before deciding on the approach to be used.

What are some of the approaches I have followed as a Lean Change Leader to get people on the Lean bandwagon?

- **Workshop:** This is a must for embarking on a Lean transformation. A Lean Change Leader should hold a leadership workshop to explain to the chief executive officer (CEO) and his leadership team what "Lean thinking" is and how it can be a "lever for business improvement." I typically run a 5-hour workshop that includes lectures, discussions, and simulations. The focus is on simulation as it will help the leaders experience what Lean can do. The quality and level of discussion during the workshop will give some sense of the engagement levels that these leaders have concerning Lean. Let me clarify that, despite a workshop, leaders can still resist, and that is when one should try the other approaches to begin the engagement.
- **Logic:** In this approach, the Lean Change Leader explains the reason why change is important and how Lean can be an approach suitable for business improvement. To convince people, he uses data, facts, and the cost of not embarking on the change. The major focus is not the anticipated benefits of the Lean efforts as shown through facts.

This typically appeals to individuals who are left-brained and are engaged through logic. For example, when a food services business was going through rough weather regarding profitability, the Lean Change Leader came up with a vision and how Lean could help on this. Most of the employees who had been with the company from the very beginning could immediately connect on the importance and came on board and supported the effort.

- **Emotions:** In this approach, the Lean Change Leader creates a shared vision and urges all to join him. When he does this, he makes all get a feel of the new order of things that will come with the change. The emphasis here could also be a set of ideals and values that are dear to the team. For this he makes things very visceral for all.
- **Pilots:** In this approach, the leader shows the power of Lean by demonstrating a pilot project. He takes up a small area within the organization and shows the benefits of Lean adoption. He does this by quick execution and making sure the results of the pilot do the talking. This approach helps to engage leaders who are doubtful of what a Lean methodology could do for their work area.
- **Personal Credibility:** In this approach, the Lean Change Leader uses past achievements and successes in transformation to influence people. People come on board because they trust his credibility and know he is someone who has been there and done that. People look up to him because of his expertise. However, this can fade quickly if the Lean Change Leader is not able to deliver things successfully in the current organization.
- **Collaboration:** The Lean Change Leader jumps into a team that is solving a chronic business problem. Although this is being led by someone else and he is only a team member, he knows his experience will come in handy when discussing and debating solutions. The belief here is that the members will see his contribution and engage with him for help in the future when there is a problem.
- **Successful Deployment:** For this approach, the Lean Change Leader takes other leaders to organizations that have successfully implemented Lean and reaped benefits. The intention here is that leaders from within the company can hear firsthand from other leaders who have experienced the power of Lean and benefited from it. This also helps to solidify what the Lean Change Leader has been telling the organization.

- **Infectious Energy:** The Lean Change Leader engages others through the infectious energy and positivity that he brings to the table. People join him because they love being with him and love his charismatic leadership.
- **Rewards/Affiliation:** This is specifically for those leaders who are hankering for recognition or visibility. They come on board because they know they could receive an award or may come in contact with the CEO or be visible on his radar. This is something the Lean Change Leader has to quickly grasp before proposing what needs to be done.

Remember that the Lean Change Leader has to have deep understanding of people and what drives them. So, as I say, a Lean Change Leader should also be a psychologist. Irrespective of the level in the organization, he treats all with respect and hears them out. He knows transformations are not easy; hence, he is patient with persistence.

9

To Whom Does the Lean Change Leader Report?

Who the Lean Change Leader reports to is a critical dimension in a company's Lean journey. If a chief executive officer (CEO) is serious about Lean deployment, he will make sure the Lean Change Leader reports to him. If you are keen on finding an organization deploying Lean, this is one of the first things that you should see. It will give you an idea whether this is a CEO-driven effort or if the CEO has outsourced the deployment to someone else. Unfortunately, there are not too many organizations in which the Lean Change Leader is driven by the CEO. However, a CEO who is committed to efficiency (not cost reduction) will make sure the Lean Change Leader reports to him. Of course, this would mean the Lean Change Leader has to be a seasoned professional who can navigate the pressures and challenges of a top management team. Figure 9.1 provides examples of typically acceptable structures for facilitating enterprise deployment of Lean.

The first two structures are clearly the ones that support enterprise-wide Lean deployment. Ideally, organizations should have this in place. There are many organizations in which the Lean Change Leader reports to the head of operations (Figure 9.1c). This sort of structure at most can facilitate Lean deployment in operations, assuming that the head of operations is on board. However, this approach also works when companies first have a plan for successful deployment with operations, after which Lean is taken enterprise-wide. The deployment in operations acts as a pilot. When the effort moves to the entire enterprise, the Lean Change Leader then reports to the CEO.

Figure 9.2 shows some weird structures being adopted by organizations that should never be done by a company serious about enterprise-wide

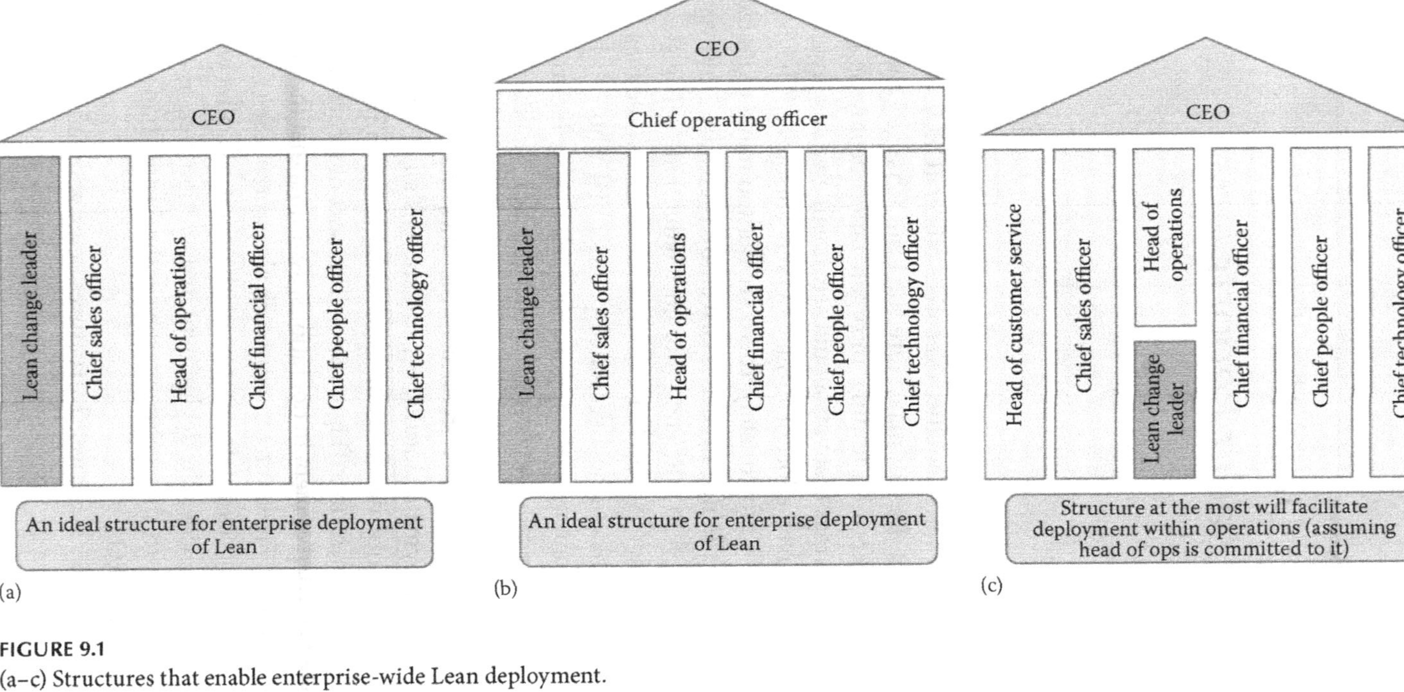

FIGURE 9.1
(a–c) Structures that enable enterprise-wide Lean deployment.

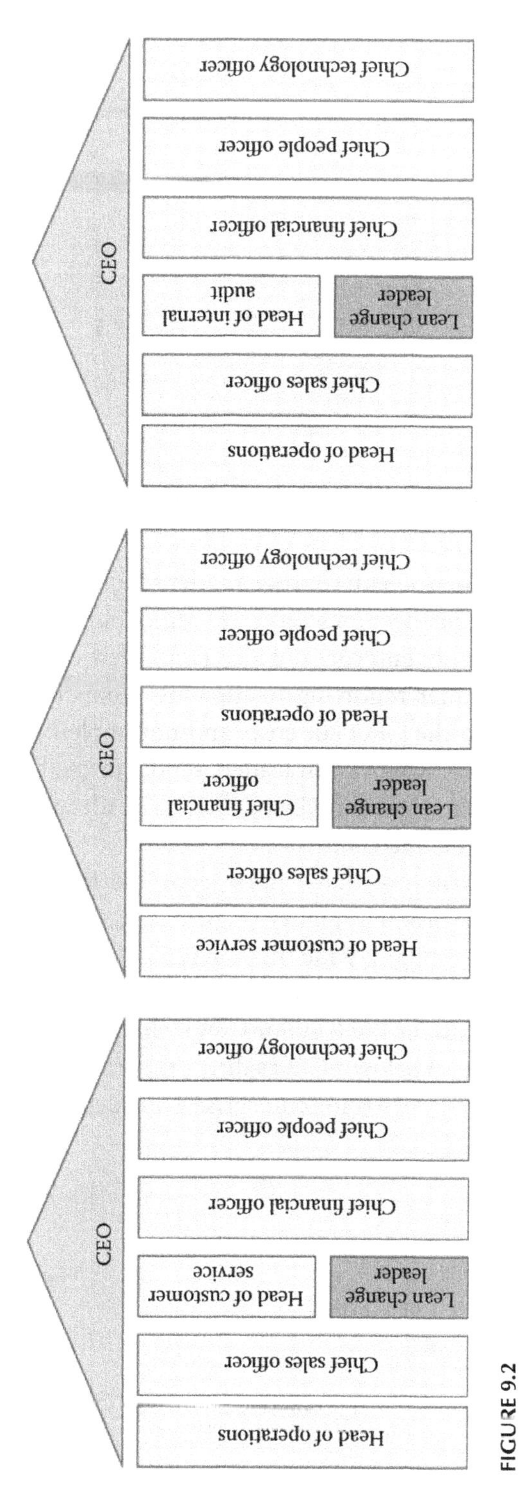

FIGURE 9.2
Structures not favorable to enterprise-wide deployment of Lean.

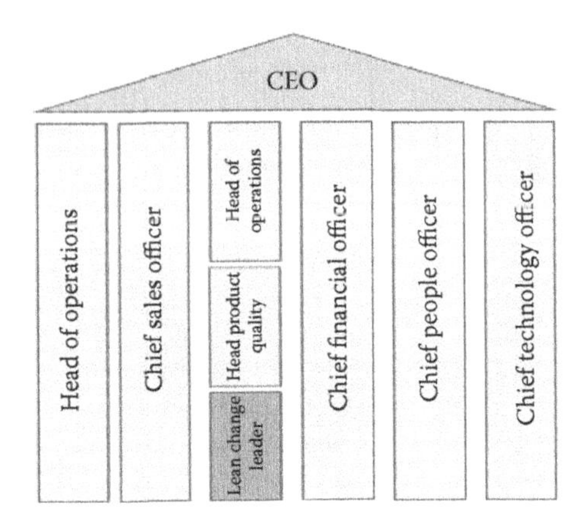

FIGURE 9.3
Lean Change Leader reports to the head of product quality in this structure.

Lean deployment. When a Lean Change Leader reports to the head of customer service, the focus becomes more on removing waste in customer service processes and not enterprise-wide Lean. A few organizations have the Lean Change Leader reporting to the chief financial officer (CFO), which is done to help the CFO cut costs and not implement Lean. Then, there are organizations where Lean leaders report to the head of internal auditing. The logic here is that the Lean Change Leader will help to build process centricity and embed controls.

There is another structure that I have seen: The Lean Change Agent reports to the head of product quality (Figure 9.3). How does one expect Lean to be deployed company-wide this way?

My personal belief is, for whatever reason, the Lean Change Leader is not able to be placed under the CEO or COO (chief operating officer). He could be placed one level below a CEO's direct report with a clear commitment that he will be supported for enterprise-wide deployment.

10

Trigger Signs of Upcoming Change by Embedding New "Ways of Working" for the Leadership Team

As an organization decides to embark on a Lean journey, it makes sense that the top management, which comprises the chief executive officer (CEO), his direct reports, and the Lean Change Leader, comes up with a set of ground rules for working together. This is because Lean thinking is about changing the culture of the organization, and this should begin from the top. What worked previously may not be acceptable when Lean deployment is under way. These ground rules or "ways of working" act as guides and make sure the interaction among the leadership team is productive. An integral part of any Lean effort is ingraining new behaviors. There is probably no better way to make the top management start practicing a set of ways of working that will lay the foundation for the Lean journey. This is really a powerful approach to start the engine of Lean thinking in the company. I have got such ways of working crafted for top management and seen their impact not only on the quality of interactions and teamwork among the management team but also on the larger organization. These shared ways of working also hold the leadership team together and help to negotiate tough challenges that come up during Lean transformation and the organization at large.

For a leadership team that has already been functioning in a certain way, this may not be an easy change. But, the Lean Change Leader has to insist and, if required, have the Lean process enforced by the CEO. What I have seen is that CEOs typically support this as it is logical and brings greater order to team functioning. So, how does the Lean Change Leader go about firming up the ways of working? He should facilitate a session and pose the following question to all (including the CEO, who is also present): *How should the leadership team function going forward to make the organization more efficient and effective?*

Then, the leaders come up with a list with the headings "Must-Haves," "Good-to-Haves," and "No-Nos" (Table 10.1).

It is important for the Lean Change Leader to avoid jargon concerning Lean and focus on the behaviors and norms that support Lean thinking. Whatever ways of working result, they should be debated by all; the Lean Change Leader should make sure they align with behaviors that build a Lean enterprise.

The initial days may require a bit of enforcement and repeated communication. I would recommend that before all meetings, the ways of working are projected on the screen so that leaders are reminded how to behave during team interaction. Also, if there was someone who faltered on any of the ways of working, other leaders should point this out. This is something no one likes as they do not want to be seen in a wrong light in front of the CEO. So, there is a large amount of peer pressure to behave differently.

For one company that embarked on a Lean journey, one of the first things that I did with the leadership team and the CEO was to come up with the ways of working shown in Table 10.2.

This had a positive impact on the leaders. Their behavior toward other leaders changed over a period of time. The CEO told me he could clearly see that the team had become more productive. Also, there was a positive impact on others. The Lean projects had yet to begin, but employees could see things changing. Meetings were starting and ending on time.

TABLE 10.1

Must-Haves, Good-to-Haves, and No-Nos for Ways of Working

Must-Haves	Good-to-Haves	No-Nos

TABLE 10.2

Ways of Working for "Prolific Business Process Outsourcing (BPO)"

Behavior/Rule	What Do We Mean?
Global BPO first	• We have a shared passion and act in the best interest of Team Global Prolific BPO. • We move beyond functional silos/look at the big picture. • We wear a Prolific hat first, then the business/functional hat.
We collaborate	• We are part of a broader network of Prolific BPO. • We demonstrate service excellence by acting in the best interest of Prolific BPO—we constructively challenge stakeholders and delight them with our servicing attitude. • We communicate clearly and in a timely manner. • We embrace change positively.
We respect trust: years to earn, seconds to break	• We demonstrate impeccable conduct; we live up to what we have committed to doing. • We respect confidentiality: Loose lips sink ships. • We are reliable and honest. • We respect each other's space, style, time, and contribution.
We call a spade a spade	• We are open and transparent. • Dissent is not bad, and there is no need to follow the herd. • We express views or opinions without reservation; we constructively challenge or counter other views and opinions. Once a decision is made, we own it.
We don't pass the buck but own it	• We take ownership and have personal accountability for decisions and the related actions. • We individually and collectively own decisions: The buck stops with Team Prolific. • We focus on end-to-end problem solving: It is Team Prolific's responsibility.

The agenda for meetings were circulated 48 hours before the meeting, and minutes were circulated within 24 hours after the meeting had happened. The employees found that the leaders who previously seldom spoke and guarded their turf were suddenly speaking to each other and collaborating. Employees also felt a new level of professionalism among the leaders. The team was the same, but the leadership team's behaviors toward each other had changed. Even if there were conflicts, the team members would know how to resolve them constructively. One of the big benefits of this is that, although a full-fledged Lean effort is yet to take off, there is nothing better than people seeing changes in the behavior among the top management team members.

11

Do You Know What Constitutes a Great Lean Team?

One of the reasons why Lean Change Leaders do not successfully achieve deployment is because they do not have good understanding of the team that needs to be put in place to get the agenda going. Building the team is not only about hiring a set of people who have Lean certification but also about the Lean Change Leader making a conscious and careful choice based on the role that the team members will play. It is this set of people that will support the Lean Change Leader's efforts going forward. It is this core team that will drive the Lean engine in the company, blessed by and under the supervision of top management. Figure 11.1 summarizes the core team that should be around a Lean Change Leader.

The core execution team for Lean comprises the Lean Change Leader, Lean Maven/Lean Expert, Lean Capability Leader, Lean Infrastructure Leader, and Lean Navigator. Let us look at each one of them next.

LEAN CHANGE LEADER

Lean Change Leaders are individuals who have the responsibility of influencing the leaders across hierarchies to adopt Lean practices for business improvement. While they are Lean Experts, they spend quite a bit of their time understanding the context concerning a Lean deployment and influence those who have a role to play in its outcome. In the first few chapters in this book, we talked about Lean Change Leaders; in Chapter 5, we discussed their traits in detail. As discussed, the Lean Change Leader should typically report to the chief executive officer (CEO) or the chief

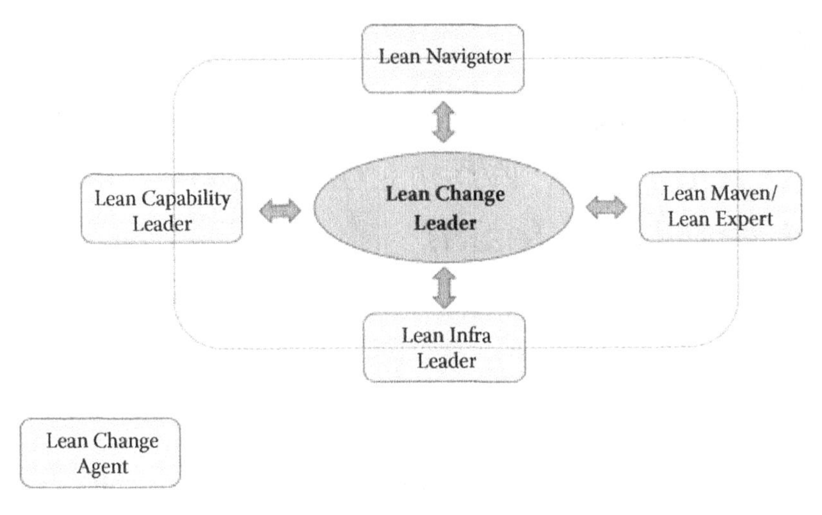

FIGURE 11.1
Core team for Lean execution.

operating officer. But, an organization could have more than one Lean Change Leader; there could be Lean Change Leaders attached to each function (Figure 11.2). This is especially true when the organization is large and it is not possible for one person to influence all functions. Some organizations call the Lean Change Agent reporting to the CEO the chief improvement officer.

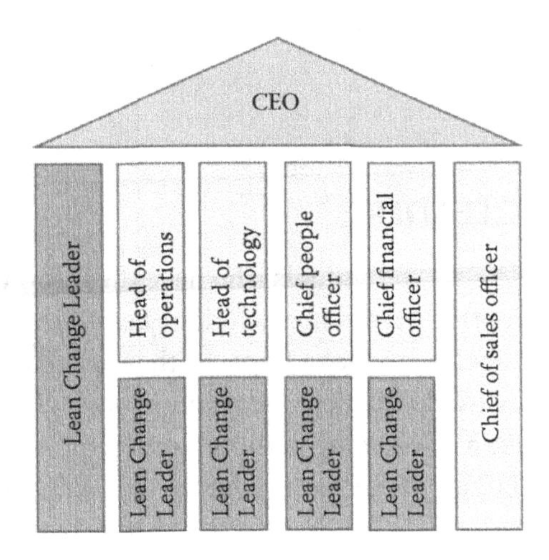

FIGURE 11.2
Lean Change Leaders in an organization.

LEAN MAVEN/LEAN EXPERT

The Lean Expert or a Lean Maven is one who has deep understanding of Lean tools and techniques and knows how they can be used for business improvement. He is an individual who has catalyzed a wide variety of Lean transformations in the past and knows what it takes to make them successful. Unlike the Lean Change Leader, most of his time goes into making transformation a reality and coaching Lean Navigators. He proactively engages those affected by a transformation and makes sure the project meets its aspirational objectives and is delivered on time and within agreed resources.

LEAN CAPABILITY LEADER

The Lean Capability Leader is one who is responsible for building capability around Lean management across the organization. He not only has deep knowledge of Lean tools and techniques but also has catalyzed many successful deployments. An ideal Lean capability leader is one who is a Lean Maven or Lean Expert who has passion for teaching others what he has learned over the years. He brings to the table not only practical knowledge but also the ability to answer questions regarding the science behind specific solutions adopted as a part of deployment. A Lean Capability Leader also has good understanding of organizational dynamics and how his capability development will have an impact on the culture of the organization. Like the Lean Change Leader, he also needs to be politically savvy as his role also requires influencing others.

LEAN INFRASTRUCTURE LEADER

The Lean Infrastructure Leader is one who builds and manages the framework of a Lean enterprise. The typical infrastructure pillar that helps sustain the Lean movement comprises metrics, communication, rewards and recognition, process management, audits, program management, management review, and best practices. The Lean infrastructure leader is the manager of all of these activities. A Lean Infrastructure Leader need not be a Lean Expert or Lean Maven but should be someone with good program management skills and the ability to manage a diverse team that supports a Lean agenda. The Lean infrastructure leader is actually the backbone of the Lean execution team.

LEAN NAVIGATOR

Lean Navigators are individuals selected from across the organization to be part of the Lean transformation journey. These high-performing individuals from the process are trained on Lean management tools and then lead Lean projects. They undergo a Lean certification program that comprises classroom sessions and successful project execution. They work full time on Lean transformation efforts for a certain period of time and then move back to their previous role or another role in the organization. What these individuals bring to the table is deep process knowledge and good change management skills to embed Lean thinking for the firm. They are considered individuals who have good leadership skills; others in the company look up to them. Given their influence, they are able to influence others to join the Lean journey.

Individuals in all five roles are also called Lean professionals as they are closely involved in the Lean transformation effort.

Although not a part of a core team for Lean deployment, we also need to have information about Lean Change Agents.

LEAN CHANGE AGENT

Any person in the company who has interest in Lean thinking is a Lean Change Agent. He has either read about Lean or has gone through an awareness program. Although not a part of the core team, he supports the agenda from outside. Some of the high-performing Lean Change Agents also become Lean Navigators by undergoing a certification program as discussed previously.

Remember, it is not that one role is better than another—all are relevant and important. Individuals are engaged in a role based on capability, experience, and leadership depth. All of these roles need to work in harmony to deliver the larger organizational outcomes.

12

A Person Keen to Be Popular Should Not Become Involved in Lean Efforts

Leading an enterprise-wide Lean movement is not an easy job. It is done by Lean Change Leaders, Lean Mavens, and Lean Navigators who drive things with a missionary zeal. They have a steely resolve to get things done despite opposition. While these individuals have respect for people, what is paramount for them is the larger purpose of improving the performance of the organization. If you are someone who believes in maintaining a workplace relationship, this role may not be for you. Remember, in a Lean professional's life, there could be many conflicting situations if the Lean professional needs to do things that might cause tension and make those around him upset. The Lean professional cannot avoid those occasions or put them down to avoid spoiling a relationship. Or, there could be times when you have to call out or facilitate decisions that disturb people's current state. All of these create a lot of disruption in relationships. There are those who would say that they believe in influencing things by being friendly. But, when you want to get things done, because of your personal relationship, more often than not you will struggle as a Lean professional. You will find it difficult to make the required tough calls as your friendship will be in the way—and you will turn out to be an ineffective Lean professional.

I know of a Lean Change Leader who joined a food service company. He was not only technically sound on tools and techniques of Lean but also had led successful transformations in the past. He was also politically savvy and could understand the organizational terrain regarding who was with him and who was not. Given his charismatic personality, he could always connect with all in the company. Within 6 months of his appointment in this role, the entire top management and many in middle management and on the shop floor became his admirer. As a great speaker, he could also motivate a team, who would become captivated by his words. Whether it was security personnel at the gate, the janitor, customer service executives, or a middle-level manager, he spoke to all. Whenever he saw them, he smiled at them and stopped to have a quick word. Somewhere in the attention and adulation, he forgot that he had been hired to facilitate the improvement of organizational performance and that this was his prime objective. He started believing that while organizational performance improvement was his goal, this could never be done at the cost of the relationships and popularity he had garnered over the past few months.

The transformation work that he took up was to improve sales force productivity. Given his past experience, he worked with the project team and came up with solutions that could improve the productivity of the sales team by 45%. One of the outcomes of this would be to release from their positions 70 existing salespeople. Because this had a huge impact on people and could disrupt the sales team, the Lean Change Leader thought of first presenting it to the head of sales (who had become his close friend) so that there would be no confusion between them later, especially when the outcomes were presented to the chief executive officer (CEO). However, the head of sales said he could not make these recommendations as it would disturb his team. When the Lean Change Agent was adamant, the head of sales told him that he could not put his friends in trouble. The Lean Change Agent, who was conscious about maintaining the relationship more than what was beneficial for the company, gave up. Without this decision to release 70 salespeople, the project could not have achieved its desired results. As a result, he could not present the final outcome to anyone.

The Lean Navigators also felt aghast as their efforts had gone down the drain. They actually felt cheated. What the Lean Change Leader ideally should have done was to present the findings to the leadership team; he should have talked about redeploying these people by helping them attain new skills. Instead, his friendship got in the way.

There were similar issues with two other transformation projects he had started, in human resources and operations. In each of these, there were tough calls that had to be taken but were not taken because the Lean Change Leader was too friendly with both the head of human resources and head of operations—the affected functional leader. Somehow, he would become unsettled whenever there were such calls to be made as he thought it would affect his hard-earned popularity. After 24 months, when he had nothing much to show as performance outcomes other than some

Ten Commandments for Lean Professionals

1. Your purpose for existence is organizational performance improvement.
2. You are not in the organization to make friends.
3. You will have a good working relationship with all, including those you may not like.
4. Your decisions are not influenced by how much you are liked or the chemistry you share.
5. Your aim is to have a trusted partnership with one and all for business performance improvement.
6. You develop a caring/organizational relationship but don't sustain it at the cost of organizational outcomes.
7. You are going to treat all employees with respect.
8. You are friendly with all yet maintain a distance with your colleagues.
9. You don't shy from making tough calls even if it disturbs those around you.
10. You call a spade a spade.

FIGURE 12.1
Ten commandments for Lean professionals.

small projects here and there, the same leaders with whom he had been friendly told the CEO that they were not seeing any benefit from Lean. What happened next? He was finally told to put in his papers.

The key lesson for the Lean professional here is that if you think you can be more effective by being popular and friendly, you are wrong. Lean transformations happen when Lean professionals do what is right for the company instead of being friendly and popular with all.

I have put together a set of 10 rules I have followed and believe all Lean professionals should follow. I call them the 10 commandments of Lean professionals (Figure 12.1).

Remember that being a Lean professional is not for the fainthearted. This professional needs to focus on organizational improvement and to stay rock solid against the pressures that are in the way.

13

It Makes Sense to Define Lean Differently

We all know many of the concepts that we use in Lean manufacturing today were nurtured by the Toyota Motor Company. Over the years, Lean manufacturing as we know it today has been defined by many thought leaders and practitioners. The following provides a peek into some of the definitions prevalent in the market:

Taiichi Ohno's succinct definition of the Toyota Production System, (1999):

> In Toyota Production System, all we are doing is looking at the time from the moment the customer gives us an order to the point when we collect the cash. And we are reducing that time line by removing the non-value wastes.

James Womack and Daniel T. Jones (*Lean Thinking—Banish Waste and Create Wealth in Your Corporation*, 1996):

> Lean thinking provides a way to do more and more with less and less—less human effort, less equipment, less time, and less space—while coming closer and closer to providing customers with exactly what they want.

Chet Marchwinski and John Shook (*Lean Lexicon*, 2003):

> Lean Production is a business system for organizing and managing product development, operations, suppliers, and customer relations that requires less human effort, less effort and less space, less capital and less time to make products with fewer defects to precise customer desires, compared with the previous system of mass production.

Wikipedia:

> **Lean manufacturing** or **Lean production**, often simply "**Lean**," is a systematic method for the elimination of waste ("Muda") within a manufacturing process. Lean also takes into account waste created through overburden ("Muri") and waste created through unevenness in work loads ("Muda").

US Environment Protection Agency (www.epa.gov):

> Lean manufacturing is a business model and collection of tactical methods that emphasize eliminating non-value added activities (waste) while delivering quality products on time at least cost with greater efficiency.

Since its application began in manufacturing, many people have viewed Lean as a purely manufacturing business philosophy. They could not be more incorrect. It is essentially a way of thinking about what you produce or provide to your customer and how you can do it more efficiently, at lower cost, and more profitably. While the underlying philosophy and principles remain the same, when one looks at enterprise-wide deployment, it makes sense to define it differently. I have used this successfully in services organizations. The definitions are as follows:

Strategic definition when the focus is the entire enterprise:

> Lean is a management philosophy that endeavors to improve the performance of an enterprise by focusing on elements that do not add value in the organizational system that is made up of leadership strategy, people, process, partners, policies, etc.
>
> An adoption of Lean should make an organization effective, efficient and adaptable, and it should be able to deliver product or service to the customers when they need it, at a desired price with an expected service experience.

Tactical definition when the focus is the entire enterprise:

> Lean is a process optimization methodology that focuses on improving the effectiveness and efficiency of a process by eliminating activities that do not add value to the customers and the product.

At a strategic level, a Lean implementation should ideally result in

- Better customer experience
- Enhanced revenue
- Reduction in cost
- Reduction in business complexities
- Added simplifications
- Improvement in productivity
- Reduced cost of business acquisition
- Better people engagement
- Reduced operational risk
- Improved profitability
- Enhanced customer loyalty

The truth is that most managers are unable to have an impact on even four of these areas. This is primarily because systems, like people, become used to a certain way of functioning. We forget to constantly evaluate them to understand if they are working for the organization and the customer. Lean helps us focus on these issues and evaluate our present systems to streamline them for increased efficiency.

At a tactical level, the implementation of Lean to a process will result in the following:

- Reduction in cycle time (time taken to complete a business activity)
- Reduction in waiting time
- Reduction in query time (time taken to resolve a query)
- Reduction in lead time (customer-to-customer time)
- Reduction in throughput time (time to complete the entire process)
- Reduction in defect levels in product or service
- Flawless processing of goods or services

My two definitions use a language that leaders in a service industry can understand easily; they cannot resist adoption of Lean because of words that they do not understand.

14

How Aligned Is the Top Management on Organizational Outcomes?

I was conducting a Lean alignment leadership session for a chicken-processing company. The company produced a well-known marketplace brand and had ruled the marketplace as there had been no competition. The company had been profitable and was known for high shareholder return. However, the chief executive officer (CEO) believed that the organization could do much more to make the business efficient. This was especially because there were newer competitors on the horizon waiting to enter the fray. He had heard about Lean thinking and reached out to me to lead a session for his leadership team before he could formally decide if he should embark on a formal Lean journey.

The leadership team comprised vice presidents of processing, marketing, engineering, product development, human resources (HR), and strategy. Before I began the session, I asked those present to write on a Post-it their strategic priorities for the next 12–18 months. The outcomes are shown in Figure 14.1.

The leaders wrote goals that were completely divergent. Does it surprise you? Well, it is a story of quite a few organizations. These organizations may be profitable, doing well, yet the top management is not aligned on the Strategic Objectives. Just imagine how much more a company could have achieved had the entire leadership been aligned on the goals. This is also explained through Figure 14.2. There are two options. A team (Figure 14.2a) that is pulling in all directions can move forward but its overall force will be much less. A team aligned in one direction (Figure 14.2b), with all focusing on common objectives, will be a powerful force difficult for market forces to stop.

As a part of preparation for the transformation journey, a Lean Change Leader should ensure the top management is aligned on the Strategic

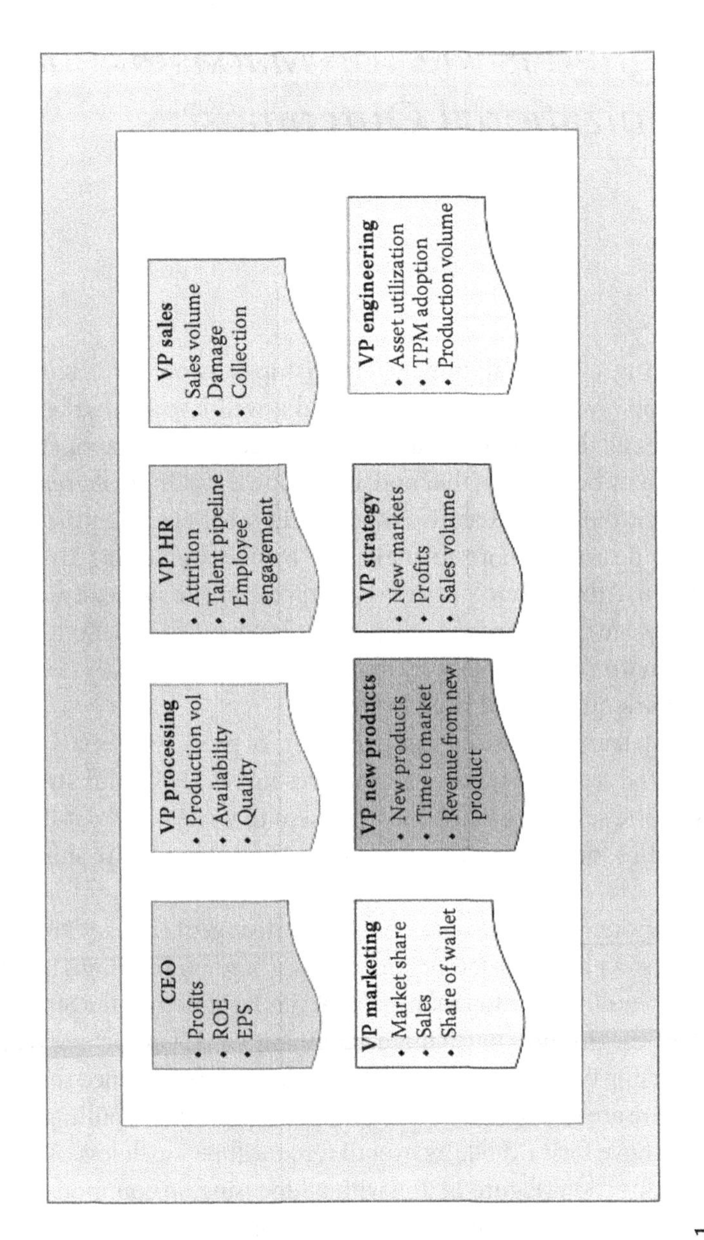

FIGURE 14.1
Lists of Strategic Objectives according to top management of a chicken-processing company. EPS, earnings per share; ROE, return on equity; TPM, total productive maintenance.

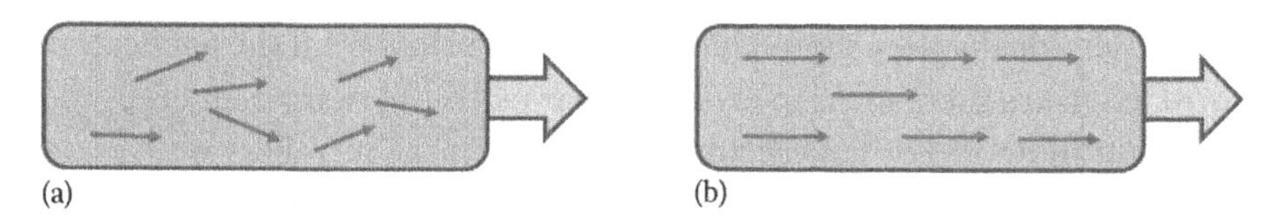

FIGURE 14.2
How aligned is the top management? (a) Leaders are focused in all directions. (b) Leaders are aligned in one direction.

Objectives. This is extremely important. I have seen Lean transformations failing not because the process was optimized correctly but because the leaders at the top were not aligned. A basic prerequisite for a successful Lean transformation is for leaders to be in sync on the organizational goals. Let me give you an example.

Symphony Finance Company embarked on a journey of Lean transformation. The first project that they decided to work on was "operational transformation" of their operations shop. The top leadership thought it was a waste to spend time on alignment on "strategic priorities" and instead got into action straightaway. The CEO said that he is a man of action and does not want his top team to spend on alignment of "strategic objectives" during midyear. This is something they will do during the next financial year when the Lean Change Leader could provide his inputs. The Lean Change Leader couldn't really influence the CEO and leaders that alignment on strategic priorities is a must before embarking on a process transformation effort. As a part of this, they embarked on improving the lead time of account opening process for a savings account. There were a large number of complaints on the long time it has taken to process these account-opening applications. Customers were peeved that they often did not know the status of their application, and when the account did get activated, it had issues of wrong personal details such as incorrect name, incorrect address, etc. The project team did a deep dive of the process and realized that one of the reasons among many other issues for long lead times was customers not providing personal details in a single go. The sales team that got these account-opening forms did not make the application requirements clear to customers. They were always in a haste to get more number of account-opening forms. The general approach that they followed was to take the relevant copies of KYC* documents and use them to fill the application

* KYC is know your customer. This is the process of verifying the identity of clients by verifying documents such as passport, drivers license, etc.

form later. Ideally they should have made the customer fill the application form to ensure there were no errors. However, this took time and they did not want to do it. For new accounts, the sales team members were mandated to meet a certain number of potential customers. And this was quite an ambitious number, which could not be achieved if they spent a long time with one customer. The problem was that at the end of the day, the sales team members filled the application form on behalf of the customers. And this was done in a hurry as it was at the end of the day when they were tired and also had to file their daily sales reports.

After having done the detailed process study, the project team probed further and found that there was a mismatch in what was important for the sales team versus what was important for the operations team that managed the back-office. The incentives of sales team members were based on the number of accounts that they got every day, while the operation team's incentive was based on the productivity or the number of account-opening forms that they processed every day. The operations team just wanted to push as many forms as possible, which sometime led to them missing out on scrutinizing the forms carefully.

Clearly both their focus areas were not aligned to what the customers wanted. Ideally both of the teams should have focused on the quality, and the quantity of accounts opened within the stipulated time and without any errors. When the project team probed further, it found that the root cause of this was because the CEO had asked the chief sales officer to increase the number of accounts at any cost, which was required to show the investors that the finance company has been performing. The CEO had asked the chief operating officer (COO) to reduce costs. Hence, the chief sales officer started tracking the number of forms sourced by the sales team every day and pushed them with unrealistic daily targets. The COO focused on productivity to ensure the number of people used for processing forms was less and helped to keep costs under control.

If the company had adopted the right approach, it would have first aligned on the strategic goals and this would have resolved this issue over a period of time. Instead, the project team embarked on looking at the process when the prime reason was strategic alignment.

This is the reason why the Lean Change Leader should engage with the CEO and top management to force the alignment on the same.

15

Do You Know the Building Blocks of a Holistic Lean Transformation?

Organizations keen on enterprise-wide deployment of Lean cannot be successful just by focusing on a few processes. If a chief executive officer (CEO) is looking at an organizational transformation, the CEO has to look beyond processes. I have seen Lean professionals struggle because their focus has just been on getting processes right and they are not aware of what else needs to be done. From my point of view, a successful Lean transformation requires nine building blocks (Figure 15.1).

Let us look into each of them:

1. **Leadership:** Leadership is probably the most critical dimension of a Lean transformation.
2. **Strategy:** Getting the strategy right sets the tone of how Lean would be adopted by the organization. An ill-conceived strategy will deliver suboptimal outcomes. When we talk about strategy, there can be two approaches an organization can adopt. One of them, "Lean thinking," is looked at as one of the business strategies of the organization. The other could be "Lean," looked at as an enabler of business strategy. To understand the difference, look at Figure 15.2. Whichever approach an organization adopts, it is important to remember that it is defined clearly, is broken into operational components, and has its resource requirement for execution well established. It is then communicated to all relevant stakeholders with a clear measure of success. One thing we should not forget is that those involved in the execution should also be represented in the strategy creation process. Clearly, organizations keen on creating a Lean enterprise should adopt Lean as a business strategy.

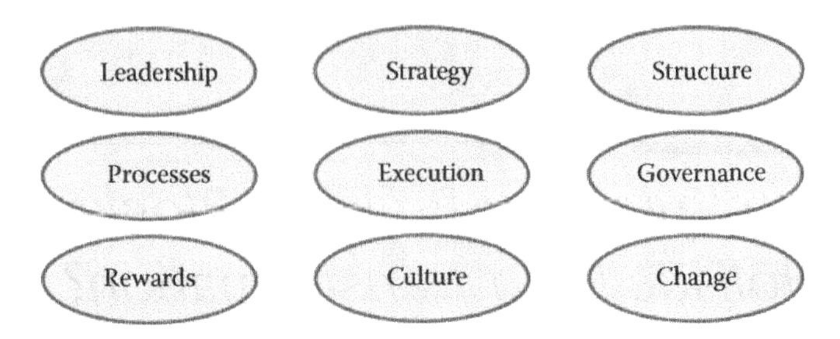

FIGURE 15.1
Building blocks of Lean transformation.

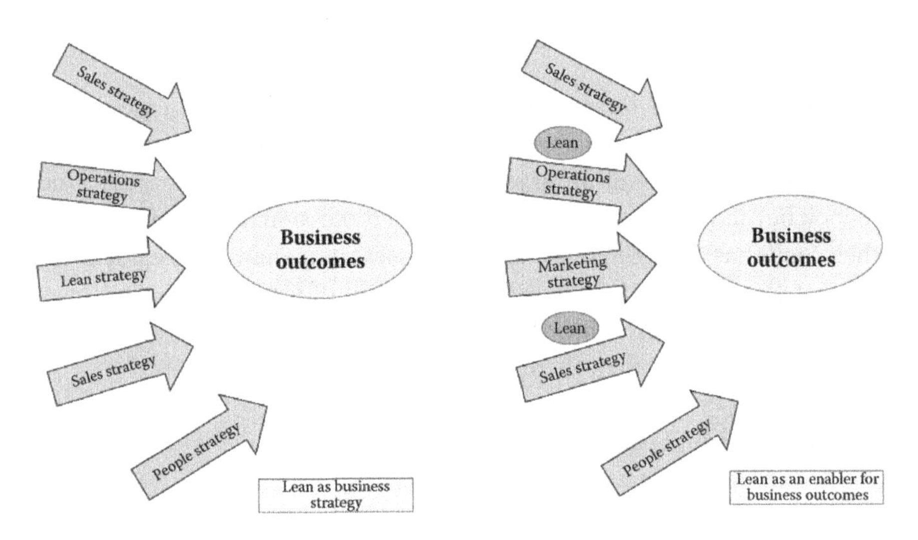

FIGURE 15.2
Lean as a strategy: two approaches.

3. **Structure:** A Lean transformation may require change in structure to ensure flawless delivery of the agreed outcomes. The type of structural change required is based on the type of strategy that is adopted. If we have decided to adopt Lean as a business strategy (Figure 15.2), the impact is enterprise-wide; hence, the structural change could have an impact on the entire organization. However, if Lean is looked at as an enabler for business outcome, the changes could be confined to the area where the deployment is under way. The choice of organizational structure among things will also look at how much it costs and what the benefit will be. How will it help in execution of the strategy, and what impact will it have on the customers? There is a perennial debate regarding which is better: a process-based organization or a

functional organization. Both have their pros and cons. The easiest thing that a Lean professional can suggest is to go for a process-based organization or an organization carved around value streams. This could embed efficiency, bringing economies of scale and reducing cost per unit. However, those against this would tell you it is not a suggested model if focus is on quick response to industry change or responding to customers. Hence, some organizations go for a hybrid structure of a pure process-based company and a functional organization. I am not suggesting one is better than the other. All that I am suggesting is that the structure needs to be decided based on the context. Whatever structure is adopted, make sure it is aligned with the Lean strategy and all affected understand why they are doing this and the benefits of it.

4. **Processes:** Processes are at the heart of Lean execution. A Lean professional will also associate Lean with process. The non-value-added efforts that we focus on create an efficient process. However, you need to remember three things concerning processes: Always focus on end-to-end processes. As a part of Lean deployment, establish owners for the firm's core processes. Never take up a process for Lean efforts if it is in transition.
5. **Execution:** A great strategy can remain a lost effort if it is not executed well. Successful Lean transformation happens by ensuring the right capabilities support the strategy and there is ownership at all levels. Everyone across levels commits to the actions central to successful execution. There is regular follow-through, and progress is measured through regular reviews. I have seen Lean execution fail because leaders had not anticipated the challenges that would come when the strategy was being devised. Also, when the duration of execution is too long, the context changes, and the assumptions that were made during the strategy session may not be relevant. Hence, I would recommend to break the deployment to a manageable 90 days. An effective feedback loop should be put in place for a framework of execution.
6. **Governance:** Successful Lean transformation requires robust governance to make sure things do not go off the track and things happen as planned. As a result, I suggest putting in place various forums to monitor the progress. Forums are steering committees for leaders to track progress. These could be at various levels, and number and size could depend on the size of the enterprise and number of

TABLE 15.1

Governance Forums for Lean Transformation

Name of Forum	Purpose of Forum	Participants	Frequency
CEO forum	For the CEO to get a sense of the overall Lean journey and the status of key initiatives/projects	CEO, direct reports, and Lean Change Leader	Quarterly
Risk forum	For the head of operations risk or CFO to ascertain if projects are impacting controls/regulatory policies	Head of ops risk/ CFO rep, project teams, Lean Maven	Monthly
Business unit/ functional forum	For business/functional leader to ascertain health of deployment in his respective business unit or function	Business leader/ functional leader, project teams, Lean Maven, Lean Change Leader	Every two months

Note: CFO rep, chief financial officer representative; ops risk, operations risk.

Lean initiatives under way. However, there are a few forums that are necessary irrespective of the type and size of the organization. The details are in Table 15.1. These forums are also used to resolve coordination or integration issues of a firm.

7. **Rewards:** Successful execution would require that people are rewarded for the right behaviors and actions that support Lean transformation. This should include both extrinsic and intrinsic rewards. Extrinsic rewards are things that are tangible and are of external value; they include things such as bonus, promotion, salary raises, and so on. Intrinsic rewards positively affect the self-worth of an individual, such as positive communication, lunch with the CEO, public appreciation, and so on. What is important is to make sure rewards are tied to both short-term and long-term outcomes of a Lean transformation.
8. **Culture:** Culture is the silent force that quietly propels or impedes a Lean transformation. Making sure the organization embeds the right elements in strands of organizational culture is key to success. Culture has an impact on the behavior of employees, which directly affects performance (Figure 15.3). The behaviors are influenced by leadership actions, rewards, and feedback.
9. **Change:** Lean transformations are change management efforts. It is important to manage them well so that they deliver the desired outcome. While there are many models that teach how a change program is managed, John Kotter's eight-step approach is widely adopted. However, one can adopt any model that one finds useful during a Lean transformation. Whichever model is adopted, it

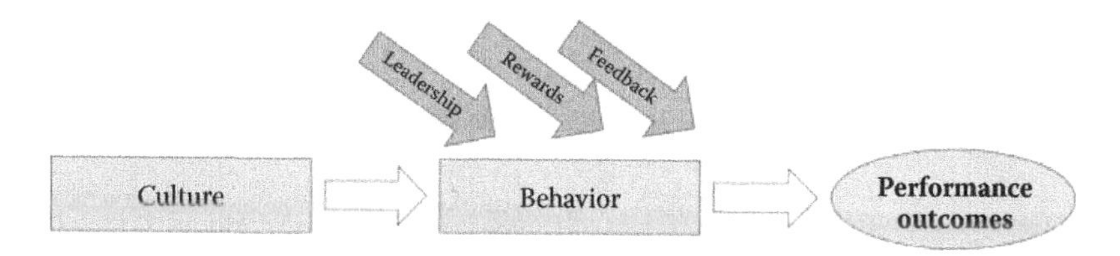

FIGURE 15.3
Culture-behavior-outcomes.

is important to remember that successful change efforts require a clear vision; a guiding coalition and a sense of urgency are critical must-haves that one cannot forget. One thing we often do not realize is that a rational change can only do so much if people are not emotionally connected with the vision for the change. It is the emotions that deliver the desired outcomes, so leaders have to connect viscerally.

16

Do Not Forget the 15Cs of Lean Transformation

Managing ambiguity is an integral part of all change programs. As discussed previously, Lean and change go hand in hand. However, to make a Lean effort, it is important to bring clarity to a host of issues so that employees do not speculate about what is happening and how they will be affected by it. I have seen Lean journeys derailed because of lack of clarity. Clarity is nothing but alignment and awareness about key issues concerning the organization and the Lean efforts that are under way. This is something the Lean Change Leader cannot do by himself but has to be owned by the entire top management. It cannot be achieved by communicating once but is something that has to be mentioned repeatedly in available forums. There are many areas on which employees need to have clarity. When communicating, leaders need to be sure they are specific. I have come up with a list of 15 major areas on which clarity is required during a Lean transformation. Figure 16.1 summarizes them, and they need to be made clear to employees during various points of a Lean journey. I call them the 15Cs of Lean transformation and have successfully used them in my Lean efforts. Just one thing that you need to keep in mind for each of these areas is that, while communicating, leaders need to be specific so that all can understand what is being conveyed.

Cs	What clarity	Details
C1	Clarity on vision	Everyone in the organization should know where the organization wants to be in a couple of years and how Lean would be contributing to it.
C2	Clarity on Lean outcomes	Clear understanding by all of the specific objectives of Lean transformation.
C3	Clarity on metrics	Clear understanding by all of how success of Lean efforts will be measured.
C4	Clarity on strategy	Clear understanding by all of the strategy and approach that will be adopted during the Lean journey.
C5	Clarity on values	Employees are aware of the organizational values and how they impact Lean transformation. They also know which behaviors are unacceptable and what could be the penalty/punishment.
C6	Clarity on Lean team	Employees should know the execution team of Lean transformation and know who the Lean Change Leaders, Lean Maven, and Lean Navigator are.
C7	Clarity on roles	Employees are aware who is doing what in the Lean transformation and what are their contributions. This is to avoid speculation.
C8	Clarity on Lean as a strategy	Employees should know the reason why Lean is the right strategy and how they can connect viscerally.
C9	Clarity on structure	Clarity on why there are structural changes and how employees would get impacted.
C10	Clarity on progress	Employees are regularly updated on progress and how Lean is impacting the business outcomes.
C11	Clarity on incentives	Employees are aware of the various rewards, both extrinsic and intrinsic, that are available in the organization.
C12	Clarity on involvement	Employees are clear on how they can get involved in the Lean journey. They know whom to get in touch with and what needs to be done.
C13	Clarity on timelines	Employees are aware of the broad timelines of the milestones in the Lean journey. Committing to the broad dates forces the leaders to stick to the deadlines as they will not like to look foolish if the dates are not met.
C14	Clarity on customer issues	Employees are aware of the customer issues that the organization has today. They should all make an effort to share and display the defective product or the complaint letter.
C15	Clarity on burning issues	Employees are aware of the chronic business issues facing the organization. This is to make sure all know what is affecting the organization and how they could contribute. This forces leaders to be transparent and bring out all issues.

FIGURE 16.1
The 15Cs of Lean transformation.

17

Let Us Not Think of Lean as a Cost-Cutting Endeavor

There are organizations today that are branding cost-cutting efforts as Lean thinking. In talking to a chief financial officer (CFO) of a leading consumer goods company, he told me that his team had embarked on a Lean effort. I was impressed to hear a CFO and not an operations leader talking about Lean. On probing, I realized what this company was doing was cost cutting and not Lean.

Unfortunately, organizations often view Lean as purely a cost-cutting tool. This is a parochial view from individuals or organizations that do not entirely comprehend what Lean thinking is and what it can do to a business. Lean is not a cost-cutting tool but a method that drives cost efficiency. This is just not an issue of semantics of words but an approach that should be followed under the contexts of both cost cutting and cost efficiency. We know that when we look at cutting costs, the objective is to bring down costs by negotiations, management mandate, or just a management call.

For example, imagine a scenario in which the travel costs in an organization have gone up almost 200% of that budgeted. The management panics and decides to do something about it. The top management team meets and makes the following decisions in an effort to arrest the spiraling costs:

- Travel costs of all sales and service managers were cut by 50% (it was found that the sales and service managers traveled the most).
- Travel per diem allowances of all employees were cut by 20%.

- All employees were mandated to travel on low-cost airlines.
- They installed a mechanism by which anybody at the middle-management level and below who travels more than two times in six months needs approval from a vice president.
- During outstation visits, employees were asked to travel by public transport such as bus, train; and not take taxis.

These are rather drastic measures for cost cutting. The costs did decrease by half, but these measures also affected business volumes drastically. The sales volume showed a downward trend, and by the year end, the business volumes had come down by almost 40%. This was because sales managers had substantially reduced traveling (given the cost pressures). They made virtually no cold calls, which they had done frequently previously to acquire new customers. Also, the customer satisfaction levels dropped radically. While the quality of products was the same, after-sale service had deteriorated. Previously when customers called with a problem, a visit by a service executive not only led to a quick resolution but also created and cemented customer loyalty. But with the cost-cutting drive, the number of customer visits had dropped; this had affected the customers' perception of service. On top of this, the unilateral policies of the company on travel alienated its employees.

This example may seem rudimentary, and you may think who in their right minds would practice Lean in this manner? There are several such examples in the business world. In a management philosophy, managers and leaders look for solutions for what they want and not what the philosophy recommends in its entirety.

If this organization was practicing Lean, it would have approached the problem differently and communicated the cost challenges to everyone. This then would have been followed by brainstorming about what in the processes contributed toward these escalating costs. The organization would then have put together teams to apply Lean in each of these processes (Figure 17.1) to make them more effective. This process would undoubtedly have been fairly time consuming. However, it would have delivered fundamentally long-lasting benefits to the organization. The teams can use Table 17.1 when examining activities in the process.

Remember, Lean is about cost leadership and entails dissection for performance enhancement. Unlike cost cutting, it is not a one-time exercise

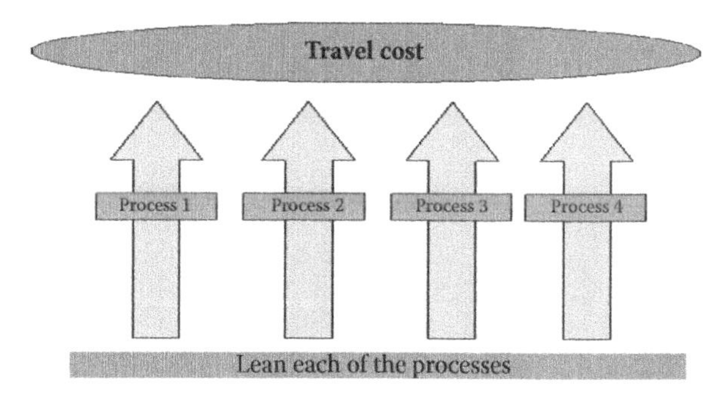

FIGURE 17.1
Lean and travel costs.

TABLE 17.1
Activities and How They Should Be Treated under Lean

Activities	Approach under Lean
Value added	Improve
Business value added (non–value added but required)	Question and improve
Non–value added	Eliminate

but something that has to be revisited at regular intervals. When you bring about cost efficiency in your businesses, it does the following:

- Makes your customers happier (because costs may go down and the benefits could be passed on to them)
- Creates an engine of continual improvement wherein teams meet on a regular basis to question processes
- Has positive buy-in of the employees (unlike cost cutting, which is often unilateral or mandated from the top)
- Brings about process thinking because whenever costs go up, teams look at processes for improvement
- Provides results that are long term and sustainable
- Increases the competitiveness of the business
- Makes the organization agile and nimble

"Cost cutting" is not Lean management. Whenever costs go up, I would suggest that management should tell its teams to evaluate the process. If your organization is using "activity-based costing," it can only add to the process. Rather, there is a need to declare a war on all those who call Lean as cost cutting and tarnish the name of Lean.

18

It Helps to Adopt a Quiver Approach in a Lean Transformation

I have seen implementation of Lean suffering because of misunderstanding on two critical issues. Some of these happen because of lack of appreciation on the part of Lean change professionals. A seasoned Lean professional would know about these and ensure everyone else also is aware of them. So what are these two issues that need to be kept in mind? Remember that Lean is just a means to an end and to institutionalize the quiver approach.

LEAN IS JUST A MEANS TO AN END

It would be short-sighted to look at Lean from a tools-and-techniques perspective. That is hardly the core of Lean. Managers must position Lean as an overarching philosophy to drive improvements in the organization. Let the overall improvement philosophy subsume the specific improvement methodologies.

The biggest mistake occurs when the improvement methodology becomes larger than the outcomes that it is supposed to produce. Remember, Lean is an approach to achieve business results. While you make your company efficient by adopting Lean tools and practices, the larger objective should be to achieve the goals that the organization sets for itself.

A Lean change professional agendum should be to achieve business outcomes. While the professional toils to achieve competence on the tools and techniques, he should never lose sight of the larger objectives. I have seen Lean professionals spend a lot of time using tools for no other reason than that they are available. Discourage teams who force-fit tools in the projects, affecting timelines and final results. It is good to achieve mastery in specific tools, but it can never be done at the expense of final results. Business leaders and Lean Mavens have to keep a close eye to make sure this does not happen.

INSTITUTIONALIZE THE QUIVER APPROACH

While Lean is implemented in your company, it would be parochial to think that a single methodology will be able to solve all problems. Remember, a one-size-fits-all strategy does not work. Based on the type of problems, specific methodologies have to be used. This is required to ensure that an optimal approach is followed to obtain the best results. What organizations need to do is to follow an integrated approach to improvement. Lean can be the overarching improvement cry, but build a quiver approach. So, what's the quiver approach? The quiver approach is about creating competencies in multiple methodologies in the organization. Let the organization build competencies in key methodologies and tools, such as Lean, Six Sigma, theory of constraints, TRIZ (theory of inventive problem solving), Poka-Yoke, process management, 5S, and so on. These are like the arrows that are kept in the quiver until they are taken out to target an enemy. The analogy is that the arrows (methodologies) are of different colors to be used for enemies (problems) of varying types or hues.

Institutionalizing the quiver approach is not easy. Organizations are like crowds and herds literally. They love single mantras and single buzzwords. Asking them to be incisive and meditative about problem solving is extremely difficult. It takes time as building competencies in multiple improvement approaches requires patience and management commitment. It would require investment of time and money, which at times management may find difficult to agree to provide.

However, business leaders serious about building a robust foundation for driving organizational excellence understand its necessity. Some leaders argue that they should build competence in the most relevant methodology. For everything else, a consultant can be hired. This is a deeply flawed approach because improvements can never be outsourced. Consultants should and can at best be used as guides to build competencies within the firm, but they never should be used to carry out

TABLE 18.1

Problem Type and Improvement Methodology

Serial Number	Problem Type	Description	Methodology
1	Ineffective output	The process is not meeting the requirements of the customers.	Six Sigma
2	Unsystematic process	Tasks are getting done, but there is a need for a structured approach to execute tasks.	Business process management system, standard work
3	Inefficient delivery	The process is laden with inefficiency, and there is the opportunity to optimize the use of resources.	Lean
4	Tribal processes	Work is getting done, but there is no defined process for the same. The knowledge of the same is confined to the process owner or a set of a few people.	Business process management system, standard work
5	Complexity inundation	The organization or process is laden with complexity.	Lean, complexity reduction
6	Nonexistent process	There is a need for a new process for a new product or service. This is related to the design of the new process.	DMADV, Lean product development
7	Workplace improvements	There are small local improvements.	7 QC Tools, Why–Why Analysis
8	Inept product or service	The product or service is not able to meet the requirements of the customer.	Lean product development, DMADV

Note: 7QC Tools: Quality Control Tools include 7 tools that are suitable for those with little training on statistics; DMADV: Define, Measure, Analyze, Design, and Verify.

improvements on behalf of the firm. This is counterproductive to never letting that problem happen again.

Table 18.1 summarizes the various types of problems and improvement methodologies that can be adopted by companies. The word *problem* refers to a situation for which there is a deficiency or gap from the desired state. The list may not be all encompassing for manufacturing but is quite comprehensive for a service business.

To summarize, make your company's methodology agnostic. Let Enterprise Lean be used as a larger improvement philosophy. Adopt the quiver approach. Build an array of competencies that helps the organization solve a wide variety of problems in a variety of contexts.

19

Let a Road Map Guide Your Deployment

Having decided to leverage Lean as a way of doing business, where do you begin? This is a question often faced by business leaders. The incorrect approach or advice can mar the deployment of this practice. Table 19.1 summarizes a broad road map for implementing Enterprise Lean in a company. To reach a future state, it could take anywhere from 2 to 5 years. While there are many books on what road map should be followed in manufacturing, this approach comes in handy when the deployment happens in a service enterprise.

Please note the steps mentioned in Table 19.1 are not sequential. Also, there could be overlap between the steps. If the future state takes a long time, I would recommend that there are well-defined milestones by which progress is seen every 90 days. Without visible change, people could lose interest in the journey or even become restless. So, the Lean Change Leader needs to make sure the milestones are defined clearly and agreed to by all. There could be other approaches for Enterprise Lean adoption in services, but I have found this approach to deliver the best result.

Often, people ask me if there are any specific outcomes that should be targeted in a Lean journey and what the sequence should be. Figure 19.1 shows an example of how a service organization targeted its outcome. There is nothing right or wrong here. However, make sure that the journey is planned in such a manner that the organization first gets its foundation and basics right before adopting complex Lean tools and techniques.

Remember that the metrics for each of the phases would be different, so refreshing the performance measures is required based on the context and maturity of implementation.

If the company has a large number of business units, the type of Lean actions will be based on the maturity and years of Lean implementation.

TABLE 19.1

Broad Road Map for Enterprise Lean

Serial Number	What	Details
1	Leadership alignment meeting	Meeting for the chief executive officer (CEO) to share with the top management that the organization is embarking on a Lean transformation journey for which it has a Lean Change Leader on board
2	"Ways of Working" workshop	Workshop facilitated by the Lean Change Leader comes up with new ways of working and is attended by the CEO and entire top management
3	Lean leadership workshop	Workshop for the top management on "What Is Lean?" and "How It Can Help in Organizational Transformation"
4	Lean pilot	First Lean deployment that acts as a proof-of-concept to establish the power of Lean; ideal scenario would have top management participating
5	Vision	Leadership to revisit the vision and see how Enterprise Lean can make a positive company
6	Lean strategy	Detailed strategy on how Lean would be adopted by the organization
7	Lean office	Create a central Lean office to program manage the deployment
8	Hire Lean professionals	Hire key Lean professionals comprising Lean Change Leaders, Lean Mavens, Capability Development Leader, and Lean Infrastructure Leader
9	Lean awareness	Create awareness among all employees on what Lean is, how to identify waste, and the journey that is planned
10	Lean Navigators	Selected talent from business units to be trained as Lean Navigators who would act as change agents during deployment
11	Install anchors	Install anchors that will help to sustain the gains from Lean deployment
12	Cultural enablers	Focus on elements that transform the organization to create the desired culture
13	Identify business unit or units for deployment	Decide on the business units wherein the deployment would commence
14	Value stream	Identify the value streams within business units
15	Core process identification	Identify the core processes within each of the value streams
16	Process metrics	Install metrics to ascertain the ongoing health of process performance

(*Continued*)

TABLE 19.1 (CONTINUED)

Broad Road Map for Enterprise Lean

Serial Number	What	Details
17	Governance framework	Embed a governance framework for oversight and ascertaining progress on an ongoing basis
18	Change program	Roll out a change management program to prepare all those affected by the change that is under way
19	Specific process improvement	Optimize a specific end-to-end process by eliminating activities that do not add value; use all relevant tools and techniques; transformation could include one or many large improvement projects (LIPs)
20	Value-stream improvement	Pick up a specific value stream of the organization and look at optimizing core processes by eliminating activities that do not add value
21	Structural modification	Modify organizational structure to facilitate building a Lean enterprise
22	Rewards	Create or modify rewards that enforce behaviors that embed Lean thinking in the firm
23	Progress check	Ascertain value-stream achievements against planned future-stage goals
24	Redefine aspirational state	Having made gains, may be appropriate to redefine the new level of the aspiration state

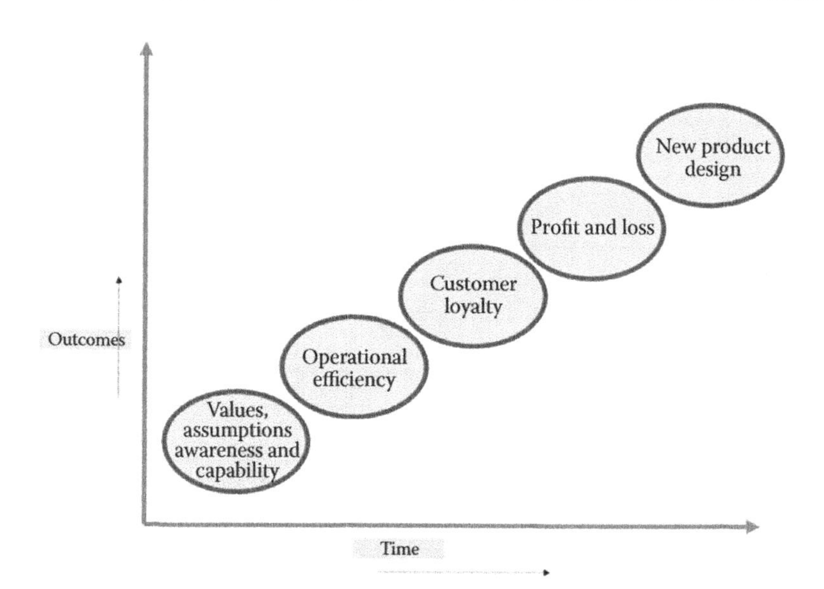

FIGURE 19.1
Outcomes from Lean during a transformation journey.

20

Observe, Observe, and Observe

Mastering the art of observation is a key aspect in Lean implementation. This chapter discusses the power of observations, which are an integral part of Ground Zero Walks.

Ground Zero Walks are time spent by leaders in the workplace where all the actions happen. *Ground zero* refers to the actual workplace. Walking, observing, and spending time at ground zero help leaders to have their "ears to the ground" to find out what is happening in the workplace.

As a part of Lean implementation, Lean Change Leaders, Lean Mavens, and other Lean professionals master the art of observation. As a matter of fact, all employees (although it is not easy) should learn the art of observation. Top management, business leaders, process owners, or any employee should make a regular visit to ground zero and observe. These observations provide insights that one cannot gather sitting in an office or air-conditioned cabin.

So, what is an observation? Observation refers to what you see in a workplace or a certain context as you walk through it or are a part of it.

Shoji Shiba, in his book *Breakthrough Management* (p. 134), talked about the three key guidelines for improving observations:

- Go to the source
- Focus on specific objects and elements
- Identify interrelationships among the objects and the elements

Observation is an art and has to be mastered under an able expert. This cannot happen through a leadership mandate. Teams have to be taught how to do it.

The following are a few objectives to accomplish through observations:

- See at the sources what is happening
- Unfurl potential problems
- Obtain a feel of what the real workplace is; see what is beyond the obvious
- Ascertain the health of Lean deployment
- Look beyond what teams present to leaders
- Connect with the teams who are at the Gemba or the workplace where the action is happening
- Understand the components that make up the context
- Get beyond data into facts
- Piece together visual images to obtain a pulse of the organization
- Understand the customer
- Understand what is working and what is not in the organization

The works of Shoji Shiba in his book *Breakthrough Management* (2006) and Gerald F. Smith in his book *Quality Problem Solving* (2000) provide great insight into how the power of observation can be used for workplace improvement.

Table 20.1 summarizes the approach I have been following to carry out an observation. Observation means using visual, audio, tactile, and olfactory senses.

Many of the points mentioned in Table 20.1 could be overlapping. Do not worry about the categories they fall into; just go ahead and observe. Often, you have to use a mix of one or more rules to reach a conclusion.

When you begin with observations, it makes immense sense to be coached under a Lean Change Leader, Lean Maven, or an expert who is adept in observation skills. Just remember that every aspect of the organization provides an opportunity for observation. If required, ask a few questions in a nonthreatening manner. Sometimes, observation is about unlearning a lot of things that you may have gathered over the years. Never allow what you know to overshadow your observations. Look at everything with a child's mind and a lot of curiosity. And, yes, be sure to listen, listen, and listen.

Remember that mastering the art of observation requires practice and takes many years. If you are serious about implementing Lean, inculcate the habit of mastering the art among all your employees across hierarchies. Of course, observation has to be followed by evaluation, analysis, and inference.

TABLE 20.1

Deb's 10 Rules for Workplace Observation

Rule	What It Is	Examples
1 Closely peruse the source	Look at the workplace closely and find out what is happening. See how things are getting done, the interactions among the teams, the workplace condition, and the mood of the workplace and teams.	During a visit to an office, look at the condition of the workplace. Have files or documents been kept in order?
2 Look at the periphery	Look at places beyond the current workplace. Find out what is happening in the environment around the workplace.	While visiting a bank, find out the cleanliness of the toilet, the tidiness in the external area around the branch, and the condition of car parking.
3 Look for issues without ownership	Look for processes or any other item that does not have ownership.	During the walk in a process shop, I found a few processes without ownership. This was despite the chief operating officer (COO) of the process shop making tall claims of process ownership.
4 Look for abnormalities	Are there issues that are abnormal and should not be there? Many times, a problem may not appear to be so and seem to be quite right.	During a walk, I found a small leak in the corner of the wall. On excavating later, it was found that there were structural issues in the building that needed to be taken care of immediately. During a walk in the kitchen of a leading hotel, I smelled something awful. On perusing, it was discovered that the kitchen had a problem with the exhaust.
5 Look at your customers	Watch how your customers experience your product or service. Sometimes, it would include meeting the toughest customers.	In a supermarket, observe how the customers are treated by the sales executives. See how conveniently the customers receive their desired product or service. Specifically, look at their facial expressions.

(*Continued*)

TABLE 20.1 (CONTINUED)

Deb's 10 Rules for Workplace Observation

Rule	What It Is	Examples
6 Look for what is missing	Unfurl what is missing in the workplace. This requires looking for what is beyond the obvious.	While facilitating a Lean project meeting, I found the teams were quiet. The reason was that the business unit had a command-and-control culture in which nobody spoke in front of their superiors.
7 Look for symptoms	Look for symptoms or signals that indicate potential problems.	While looking at a presentation of a large hospital, I found a large number of unplanned casual leaves. On perusing later, I found that the team was highly demotivated and lacked a sense of ownership for their workplace.
8 Look at people's behavior	See how people are behaving and talking to each other. This speaks a lot and is a manifestation of past and current experiences.	A process owner of a leading organization told me how proud he was that no problems had been reported to him during the last six months as the process had become highly efficient. Later, it was discovered that the culture of the company was such that reporting problems was taboo. Individuals were rebuked when they reported workplace problems; this was very anti-Lean.
9 Look for legacies and remnants of the past	See if there are remnants from the past that tell something about the company or the workplace, something that happened in the past and has a link with the present.	While walking through a front office of a hotel, I saw a large queue to check in. On probing, it came out that the information technology (IT) system used for checking in guests was slow. Later, I discovered that this property had recently been acquired by an international hotel chain, and the IT system was still inefficient and old and had not been upgraded.

(*Continued*)

TABLE 20.1 (CONTINUED)

Deb's 10 Rules for Workplace Observation

Rule	What It Is	Examples
10 Do comparisons	Compare with existing practices, standards, and data from within the company and other companies that you may have seen. This could provide opportunities for action.	In a bank back office that sent monthly account statements to customers, it was found that the company was using an expensive international courier. On comparison with another low-cost local courier, it was discovered that both took the same time to deliver the documents to the customer, and the defect levels were also the same. This comparison led to the bank changing to the local courier. This resulted in cost savings, and there were no customer issues. A little comparison can lead to great improvement.

21

Lean Need Not Necessarily Be Called Lean

In my experience, I have seen leaders who are averse to methodology. When you tell them that the organization needs a Lean intervention, they take it as another of those tools that the improvement practitioners peddle to obtain their mindshare. The leaders could question the effectiveness of earlier initiatives, such as Six Sigma, elementary problem solving (7 QC Tools), process management, and so on. This is primarily because the earlier approaches had not delivered benefits, not because the approaches were bad but because they had been deployed in an ineffective manner without leadership buy-in.

I have had chief executive officers (CEOs) telling me repeatedly that they want improvements and do not care which methodology is adopted. The erstwhile CEO of a leading investment banking firm once told her team that methods such as Lean, Six Sigma, and process management are just tools. What is important is the positive impact that it brings to the table. She was so right. Improvement methodologies are just means to an end.

This is where you need to obtain a sense of organizational thinking. If you believe that leaders are averse to specific methods of improvements, just rename the deployment of Lean management as something that suits the organization. The names that companies use could be a drive for "operational excellence," "process excellence," "performance improvement," "business improvement," and so on. At a leading bank, the Lean rollout within the branch banking channel was called "symphony."

However, what needs to be kept in mind is that the philosophies of Lean thinking should be embedded in the method. The key ones that all service companies need to keep in mind while rolling out Lean are the following:

- With anything that you do, keep in mind the impact that it will have on the company in the next 25, 50, and 100 years.
- Make your company outward focused rather than inward focused—keep the customers in mind in everything you do.
- You should have respect for people.
- Always keep an eye on the impact being made on the community.
- Make sure that the laws (regulations, etc.) of the land are adhered to at all cost.
- There should be a positive impact on all stakeholders, such as customers, shareholders, partners, and so on.
- Are you faced with a problem? You must go and evaluate the process.
- Create leaders at all levels who live the company's philosophy.
- Embed the workplaces with processes and procedures.
- Make things simple.
- Create a culture of continuous improvement, value-stream thinking, and constant reflection.
- All decisions and change should be done with the team's involvement.
- Make the company unapproachable to all types of fraudsters, money launderers, and the like.
- Ruthlessly protect the knowledge of the company.

Just follow the quiver approach discussed previously. Also, instead of calling your improvement teams Lean Mavens or Six Sigma Belts, you could call them change agents.

Remember that what is important is practicing the philosophy of Lean management and not losing sight of the primary objective by fighting over a "name." As a Lean Change Agent, you need to understand the minds of the CEO and other top leaders. Many times, they may not be overt about their dislike; however, this has to be ascertained by all means. The best thing to do is to let the name emerge from the company. You could run a contest and let an employee name the process, which could then be validated by the leadership team.

22

Service Guarantee Can Be a Good Aspiration to Have in a Lean Transformation Journey

Offering a service guarantee can be a good goal to engage a company traversing a Lean transformation journey. While this may not be the only mission of a company, it creates sufficient pressure on the company to make its processes not only effective but also efficient to meet the objectives. When you guarantee a product or service to a customer, it means that you shall compensate the customer financially if the promise is not fulfilled. This is not easy and requires relevant processes cutting across silos to function at a certain level of performance.

I have seen that service guarantees can be inspirational for driving operational excellence through Lean as they require a focus on the customer, product, and process design. So, what is a guarantee? It is an unconditional promise made to the customer that ensures that if the predefined service delivery standards are not met, the company will compensate the customer. For example, a bank promises its customers that in its branches, the teller will deliver cash in 3 minutes, and it will compensate with $5 for every occasion when the customer does not receive the money in time. Table 22.1 summarizes a few "to-dos" that we need to remember before offering a service guarantee.

The target of offering a customer guarantee is so powerful that it can integrate the whole organization toward a larger objective for customer satisfaction. And, the practices of Lean just become an enabler here. However, achieving such a target would require working on all aspects of customer delivery, such as people, process, technology, and customer understanding. A customer guarantee not only can differentiate a company but also can actually create a solid working engine for Lean management. It acts

TABLE 22.1

To-Dos for Customer Guarantees

To-Dos	Details
Select a product/ product family	Based on the market scan, customer segment, and customer needs, select a product or a product family. During the early days, go slow and do not take too many products. Channel all resources to ensure that the promise is delivered.
A shared vision	Begin with the chief executive officer (CEO), value-stream owner, or business head to create a shared vision and clearly state that a customer guarantee will be a service differentiator and a key focus area for the next few years. The top management should communicate to all the need and reason for a guarantee.
Voice of the customer	Listen to what the customer actually requires. It is needed to design the right product or service and also to obtain feedback when the product is launched. Create an ongoing feedback system using multiple methods among different customer groups. Often, organizations adopt a single methodology and believe that they have a comprehensive approach to obtain the customer's feedback. A single approach is inadequate as each has its own limitations. Based on the requirement, leverage the power of approaches such as transactional surveys, global surveys, mystery shopping, declining and lost customer surveys, focus group interviews, customer advisory panels, customer complaints or comments, employee surveys, and so on.
Value-stream map	Focus on the value stream that will be used to deliver the product or service for which the customer guarantee will be provided. Find out the wastes within each of the core processes within this value stream. This will highlight the areas for improvement and the future state that needs to be achieved for offering the guarantee.
All processes need not be perfect	Do not wait for all processes to be perfect. Focus on improving the vital few value-creating processes that have a direct impact on the customer and the proposed service guarantee. Improve these processes with the customer's expectations in mind and work toward reducing the potential failure points.
Customer service specialists	Develop competent customer service executives who are able to provide flawless service to customers. These individuals are trained not only on product or service characteristics but also on the way customers' moods and emotions are managed. Adequate doses of soft skills should be imparted to them. Empower them to make instant decisions on customer complaints or queries.

(*Continued*)

TABLE 22.1 (CONTINUED)

To-Dos for Customer Guarantees

To-Dos	Details
Make guarantees unconditional	A well-designed service guarantee should be unconditional. There should not be any exceptions or conditional qualifiers associated with it. There should not be any asterisks, ifs, or buts.
Clarity of guarantees	Ensure that your promise to the customer is clear and explicit. It should not be open to any interpretation. The promise has to be pointed and sharp, and the customer should clearly know the compensation associated with the service failure.
Attractiveness of guarantees	Make sure that the compensation associated with the guarantee is attractive enough for the customer to invoke it if dissatisfied. Do not make it so measly that the customer deems it not worth the effort.
Customer feedback process	Ensure that the process of receiving the compensation is hassle free. The process should really be Lean; that is, it should not only be fast but also give an aha! feeling to the customer.
All facets of service are a clear "no"	In the early days, do not target all aspects of service around which the guarantee needs to be threaded. You can just focus on one aspect of a service, such as lead time to deliver cash by the bank teller. However, as processes mature, all aspects of service of a product can be covered, such as the full satisfaction guarantee that is provided by a hotel.

as an arsenal that drives consumer-focused improvements, which in turn positively influence the behavior of existing, potential, and dissatisfied customers. It would require organizational leadership to take action on tacit issues of complexity management, process handoffs, process work-around, silo issues, customer-centricity, and so on. I call these tacit issues because companies often do not focus on them, but they are critical to superior customer service.

23

Getting Top Management Commitment Is Necessary but Not Sufficient

It is not difficult to understand that Lean has to be driven by top management; this was discussed previously. Without the mindshare of the top management, it may be futile to start a Lean transformation. When I talk about top management, I am referring to not only the chief executive officer (CEO) but also his direct reportees. When we talk about leadership engagement, it has to be much more than just doing lip service. I have seen many CEOs talking about all the right things in their speeches, but if you talk about investing time to find out what is going on, their calendar would suddenly be filled with meetings.

A CEO does remain busy with important things. However, if he believes that the Lean implementation is important, the CEO shall surely find time. After all, a CEO will only be busy with things that he deems important. If he believes that implementation of Lean can be delegated to the continuous improvement team, it is a wrong assumption. A CEO interested in Lean has to demonstrate a visible commitment to Lean; this can be done in many ways, some of which are as follows:

- Creating a shared vision on what the transformation of Lean would lead to
- Spending time to review progress on a regular basis
- Making a public commitment about the percentage profit improvement that will be brought about by Lean
- Making all direct reportees (of the CEO) responsible for Lean transformation in the company
- Threading outcomes of Lean transformation to the performance management scorecard of all top leaders (the CEO and his direct reportees)

- Spending time to go through a 2-day leadership program on Lean management
- Working to overcome the nonalignment at the top
- Not jettisoning his personal commitment when his chips are down
- Participating in Ground Zero Walks
- Meeting or visiting customers

Just feeling good about commencing on a Lean journey is not sufficient. What top leaders should do is to ensure that Lean is leveraged not only for bottom-line gains but also for achieving other strategic business objectives.

However, it should be noted that getting leadership commitment is necessary but not sufficient. In today's world, when the tenure of the CEO has reduced, just having CEOs' commitment and that of others in top management may be short-sighted. In such a context, what we need is a plan that ensures there is adequate engagement at the top management, middle-management, and grassroots levels.

While top management has largely a strategic role in a Lean journey, middle management makes it happen. It is imperative to have an engagement strategy to convert these individuals to agents of change. Chapter 25 discusses thoughts on middle-management engagement.

Also, the ranks below middle management, the front liners and those at the grassroots, all have to be simultaneously taken on board. This should happen at the early stage of implementation. The worst thing to happen is to leave them to read the top management's mind on the proposed transformation. If they are not engaged with the transformation, the teams can surmise concerning that there will be things such as layoffs, another agenda from top management, and so on. They need to receive communication on the larger agenda and be trained on foundations of Lean thinking. This has to be backed by supporting structures that allow teams to operate with the new behavior. It is the onus of top management to ensure that the rank and file are engaged. They cannot be treated mechanically if we expect them to add and contribute to the Lean transformation journey.

Remember, Lean has to be for the people, by the people, and of the people.

24

Creating a Sense of Urgency Is a Prerequisite for Successful Lean Deployment

Just deciding to embark on a Lean journey is not sufficient. You have to create and communicate to all concerned the necessity for the journey. Creating a sense of urgency can leverage this.

Creating a sense of urgency is about communicating the importance of Lean as the basis for business improvements to all and sundry. It is about sharing the cost of the status quo and the impact it will have on the company. The endeavor should be to paint such a picture that it creates a concern in not only the leadership but also the teams below the leaders that the change is a prerequisite for survival of the business. Go to the extent of telling them how this will affect them personally or their work. This should motivate them to take immediate action. But, remember, the change has to be rooted in something that is real and of operational need. It cannot be something abstract and esoteric. Make sure that the business case of the Lean transformation is communicated to the employees in the context that is relevant and understandable to them.

As Lou Gerstner (2002) mentioned in his book, "The *sine qua non* on any successful transformation is public acknowledgement of the existence of a crisis." Communicating a sense of urgency helps to overcome the overt resistance and is about assuring employees that they will not be blamed for their past work. Leaders at all levels have to demonstrate a sense of urgency for getting things done. They should be in a position to create images in the employees' minds of the future after the transformation. Tell everyone about the TINA factor: There is no alternative. The compelling message has to be to change or perish.

You must be wondering if business is hunky-dory and the leadership is committed, then why is there a need to create a sense of urgency? Well, this is required to ensure that all the goals set out under Lean transformation are accomplished. Creating a sense of urgency is imperative for driving any business transformation. Even in a highly successful company such as Toyota Motor Company, creating a sense of urgency is a regular phenomenon. In the book *Extreme Toyota—Radical Contradictions that Drive Success at the World's Best Manufacturer*, authors Emi Osono, Norihiko Shimizu, and Hirotaka Takeuchi mentioned the following: "'You have to put your life on the line in order to make something good,' said Watanabe a few months after becoming president. 'If you compromise in the process, nothing good will come of it. If you listen to this person's and that person's opinion, your spiky horns get dull. He is going to extremes in suggesting that employees put their lives on the line or keep their horns sharpened, but such remarks resonate within the organization.'" Watanabe seized every opportunity to instill an atmosphere of urgency.

You may be the best-performing company, but remember that only those paranoid of the status quo can survive. The following is a list of must-dos required for creating a sense of urgency before you start Lean transformation efforts:

- Make sure that the message is sincere and does not look contrived and fake.
- At least 80% of the leadership team (chief executive officer [CEO] and his direct reports) should believe that there is a crisis-like situation and the Lean journey is a must for survival.
- Use all available opportunity to communicate the importance and urgency for Lean adoption.
- Communicate to the entire organization the core reasons driving it to adopt Lean.
- Share all facts so that the employees clearly know the reason for the change.
- Use various channels to communicate the need for change, such as video, town hall meetings, and presentations by executives, posters, metrics, and targets.
- Achieve a critical mass of employees who will lead the change through Lean; 30% of your employees should be in this segment.
- Try to stir the emotions of the employees so that they can be convinced.

- Shun the mindset of "we know it all." If you see this in your employees, the CEO and other leaders should break the backbone of this mindset.

Remember, just creating a sense of urgency once during the start of the Lean journey is not sufficient. It has to be done again and again. It has to be anchored on a solid business case but driven by emotions.

25

Do Not Forget to Include Those Below Top Management

In a Lean transformation, special attention should be given to ensure that middle management is taken on board adequately. This is required because the middle-management layer often does not receive complete attention.

Organizations spend a lot of time engaging the leadership team by conducting alignment workshops, and preparing for them is a part of management reviews. Teams on the shop floor, individuals working on the process, and front liners are taken through awareness programs and encouraged to take up projects. However, little attention is paid to engage middle management. This makes them disinterested in the transformation. The thought that bothers them is, why should we be a part of a movement that has been launched without involving us? The feeling that bothers them is that this is another of those "flavor-of-the-month" type of initiatives that come and go. What further bothers them is that it would increase their workload.

In an article "Who Wants to Be a Middle Manager?" published in *USA Today*, author Stephanie Armour mentioned the following: "Middle-management jobs have become more demanding. Technology means middle managers have to do more multitasking and are expected to be accessible to their staffs, a Herculean challenge in the age of globalization. Employees may be spread across the globe, and a manager may have to get up at 3 a.m. to take a call from an employee in another country."

Given these challenges, getting middle management on board requires extra effort and senior management's attention. Remember that middle management with partial mindshare results in a stillborn engagement (see Figure 46.1).

In today's world, when the shelf life of chief executive officers has diminished, full engagement of middle management is a must for sustaining the gains from Lean. As a matter of fact, one of the true proxies of successful Lean transformation is when middle management engages in driving and making the Lean transformation work. Top management will set the vision, but it is middle management that makes it happen. Middle management will communicate with people on the grassroots and front-line levels on the impact of change. It is imperative that a program be designed within the larger transformation agenda to engage middle management. The following are some of the must-dos:

- There should be an alignment session with middle management to specify the relevance of Lean and how it would create a high-performance business.
- Make each person in middle management learn how to appreciate and identify waste.
- Familiarize middle management with both the strategic and the tactical aspects of Lean.
- Look for individuals with high influencing abilities and passion for change. Give them ownership to drive Lean in their respective work areas.
- Give ownership of a few critical projects to leaders in middle management. However, these projects should have sponsorship of top management or the link may be lost.
- Use middle management as the communicator and translator of management's thought or perspective on the transformation.
- Clearly communicate that this is just not another flavor of the month.
- Seek the feedback of middle management to design and cocreate the various programs in the Lean journey.
- Capture middle management voices and expectations regarding their perspective on Lean.
- Top management should regularly ask what support is required.
- Organize visits to companies who have managed good traction on Lean.
- Provide rewards for good performance.
- Clearly differentiate between the compensation of middle management and that of front liners.
- Involve middle management in the change process—the more they come on board, the easier the acceptance and adoption will be.
- Different people have different concerns, so try to address all senses if possible and, if required, apply various modes of communication.

One of the mechanisms to ascertain the engagement of middle managers is to find their engagement levels on a regular basis. Administer the Lean engagement instrument every quarter and establish an engagement index. If you do not see a positive trend in the index, it would mean that something needs to be done or the current engagement process is inadequate.

Remember, middle managers do have a lot of information and directly manage the front liners. Middle managers are the ones who deploy strategy and help achieve the short-term goals of the company. You can never progress on a Lean journey without taking them on board.

26

Is Lean Applicable in Your Organization?

You have heard of Lean thinking but are not sure if it is something relevant to your company. You have been told about the benefits but believe that this is another of those methodologies that may not address your business plans. People tell you that Lean is only relevant to the automobile sector, and you have doubts about its applicability in the service sector.

Under such circumstances, it may be a good idea to use the questionnaire given in Table 26.1, which actually tells you if Lean as a practice is relevant to your company. Answer all the questions. Should you answer yes to any of them, it would indicate that Lean is relevant to your business.

While the questionnaire is not a solution to your problems, it will definitely give you an idea if Lean is the right thing for your business. However, it should be noted that such questionnaires are directional, and detailed assessment should be done before getting into a Lean engagement.

TABLE 26.1

Deb's Lean Opportunity Questionnaire

Serial Number	Questions	Yes/No
1	Are there chronic customer issues about which customers continually complain?	
2	Is the number of employees increasing in proportion to the increase in sales volume?	
3	Are there core processes that do not have ownership?	
4	Do you see that the capabilities of team members are not being used to the fullest?	
5	Do you see a large number of Management Information System (MIS) reports being generated that no one looks at?	
6	Do you see teams spending a lot of time firefighting?	
7	Do you see a lot of silo issues and teams not talking among themselves?	
8	Are there follow-up teams whose job is to coordinate and get things done among departments?	
9	Are error or defect levels high in processes?	
10	Are there critical issues that do not have ownership so they are not receiving attention?	
11	Do processes become automated whenever there are bottlenecks and hurdles?	
12	Are there a large number of auditors and inspectors to ensure no defects occur in the process?	
13	Do teams get instituted to spend time on process work-arounds?	
14	Are there teams to manage and correct customer problems?	
15	Is a lot of staff member time spent on urgent issues, customer complaints, and process corrections?	
16	Are there teams dedicated to exception management?	
17	Are there units within the business that do not have well-defined processes?	
18	Are there issues pertaining to consistency of delivery?	
19	Are there issues of managing scale and growing volumes?	
20	Are there processes or business operations that have become complex over a period of time?	

(Continued)

TABLE 26.1 (CONTINUED)

Deb's Lean Opportunity Questionnaire

Serial Number	Questions	Yes/No
21	Are there products or brands that have been added over a period of time and are not profitable today?	
22	Do you have nonprofitable customers who cannot be weeded out today?	
23	Are the efficiency ratios of your business going up? (An efficiency ratio in case of banking would be "cost/income" ratios.)	
24	Are you finding it difficult to manage a large number of products or service variants?	
25	Is there a proliferation of channels and products that you are finding difficult to manage?	
26	Is there a dire need to create capacity within your business to manage future needs?	
27	Is your business finding it difficult to survive the cost and competitive pressures of the marketplace?	
28	Is there a need to simplify your business processes and structures?	
29	Is there a need to have flexible business processes and systems?	

27

Service Processes Are Quite Different from Those That One Sees in Manufacturing

Lean can be applied to services, but it is not as easy as it seems to be. Given the complex structure of service processes, they provide particular challenges that a Lean change agent should be aware of. Many of these challenges arise because service processes function in a different way.

SERVICE PROCESSES MAY NOT BE VISIBLE

Manufacturing processes are visible, and we see input resources converted to product output. However, this is not the case with service processes. Inputs and resources that are used in service processes may not always be tangible. Hence, it is critical to define and identify the basic contours of the process before it is taken up for Lean treatment. This is especially true for service processes that are not very old, and no effort has been taken to document them for future reference. The problems further escalate when technology is used as an enabler in service processes. For a rookie to understand such a process is not easy. This is where tools such as SIPOC-R (supplier, input, process, output, customer, and requirement) are a great help.

SERVICE PROCESSES ARE MANPOWER INTENSIVE

One of the key resources used in service processes is people. Their contribution is the key toward meeting the objectives of the process. This is not easy as it requires committed and motivated people to work efficiently and to meet larger objectives. Tools and techniques can improve processes; however, they can be sustained only by reorienting the behavior of teams so that they proactively look for wastes and work toward eliminating them. This is where a Lean intervention has to move from mere process improvements to making Lean a part of the organizational culture.

SERVICE PROCESSES MAY NOT HAVE METRICS

Tracking the performance of service processes is still a relatively new idea in many industries. Most often, performance evaluation of processes is reactive, based on quality of outcomes or customer feedback/complaints. Such processes lack effective metrics to continuously track the performance and health of processes. In such a situation, we first need to find a baseline for the process, to understand its state, before taking it up for improvement.

IDENTIFYING WASTE MAY NOT BE EASY

Given the invisible nature of service processes, identifying wastes in them is not easy. Often, symptoms of wastes and inefficiencies may not be as overt as in manufacturing companies. There could be lots of e-mails, data, information, papers, documents, or redundant procedures that are visible only to highly trained eyes. One has to learn to spot these not-so-obvious wastes. This is where value-stream mapping (approach to identify wastes in a process) is a great help; this is discussed in greater detail in later lessons.

OUTPUT OF CUSTOMER-FACING PROCESSES CANNOT BE CORRECTED

In a manufacturing context, poor product quality can often be rectified before it reaches the customer. In services, however, especially for the customer-facing processes, the output is delivered to the customer instantly, with hardly any opportunity for corrections. So, service processes have to be impeccable and virtually error free. If errors occur, there has to be a solid service recovery process so that dissatisfaction of the customer can be converted into a business opportunity. Also, for a customer-facing process, the process itself is the product as the customer experiences not only the product but also the process that is used to deliver the product. It is the experience that matters and is an integral part of the service product. Every service product requirement is difficult and should always include the emotional component. This becomes critical as the Lean improvements have to be targeted toward meeting the objectives of the customer. Service is after all about people.

LAYOUT DESIGN OBJECTIVES IN MANUFACTURING AND SERVICE SETUP ARE QUITE DIFFERENT

Lean is all about creating flow, and this is where we endeavor to create a suitable layout. The objectives of creating a process layout in both manufacturing and service companies are quite different. The primary objective of layouts in manufacturing is the reduction in transport time. In a services setup, the primary objectives would be productivity enhancement of employees, reduction of communication cost (such as in back-office processing), and customer convenience (such as in a bank).

28

Do People Know Why the Organization Is Embarking on a Lean Journey?

Implementing Lean is about driving a fundamental change in the way organizations operate. Like all change initiatives, it is imperative that we communicate to all concerned the reasons and benefits of implementing Lean. Everyone in the company should know the larger intent and objective with which it is driving this transformation. A joint team with representatives from top management, middle management, and front liners should take the ownership to communicate the answers to the following half-dozen questions to everyone in the organization:

1. What is Lean?
2. Why is there a need to adopt Lean now?
3. How does Lean thinking contribute toward achieving the larger vision of the organization?
4. What are the key business objectives that we plan to accomplish through Lean?
5. How would it benefit each one of us?
6. How can each one of us (employees) contribute in the transformation?

Remember, this is not an easy task, especially if the organization is large and scattered across geographies. Someone from the senior management team must take ownership of making this happen. Each of these communication meetings should also be attended by constituents from top management, middle management, and front liners; if there is a union, the representative from the union should be a part of these meetings. This means that before starting the communication exercise, the management should have complete alignment with union leaders.

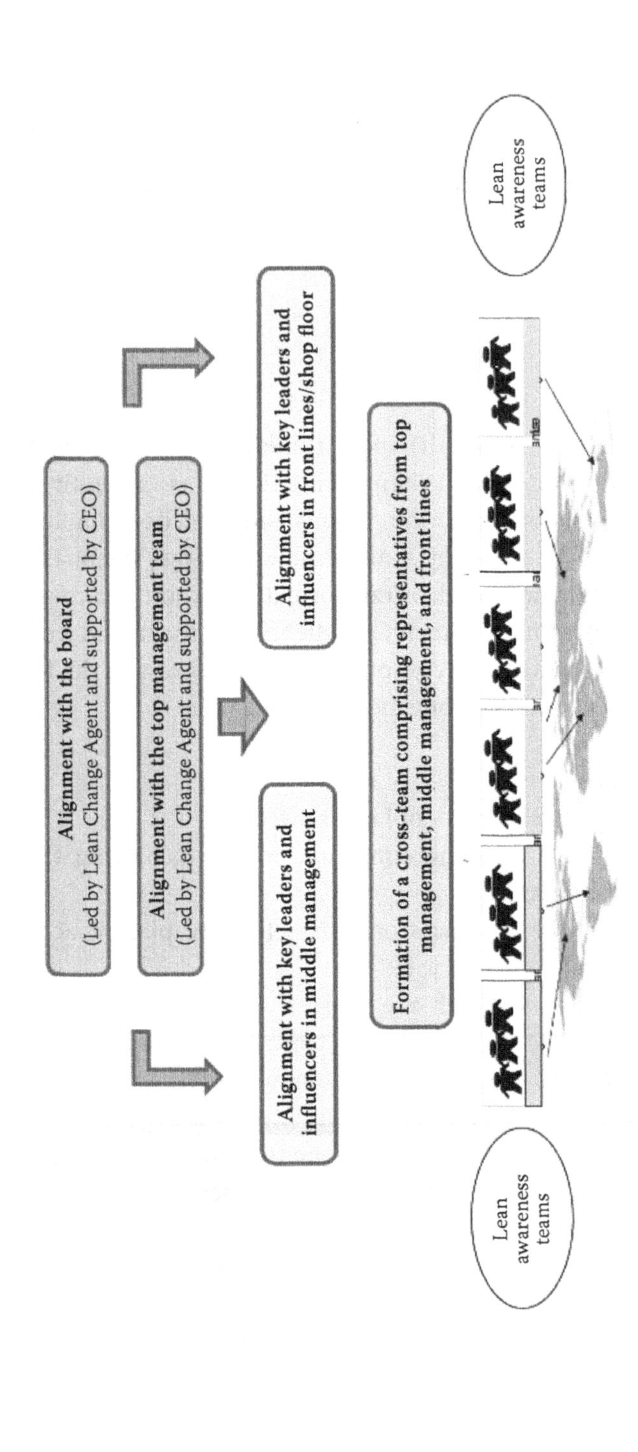

FIGURE 28.1
An approach for alignment in a global enterprise.

However, it is important to remember that awareness without immediate action is of no use. If there is a long gap between the communication and action, teams can lose sight of what the goal is. It is recommended that the actions on Lean should commence within two weeks of the alignment session.

Figure 28.1 summarizes the approach to align a proposed Lean transformation within an organization. This begins with the alignment with the board, followed by a download to the top management. Both of these are led by the Lean Change Agent and supported by the CEO. Sometimes, organizations may also take a call to invite an outside Lean expert to do these two sessions, but my preference is a Lean Change Agent who has been hired to catalyze the transformation within the enterprise.

In both these alignment sessions, the CEO should clearly demonstrate that the initiative has his mindshare and attention. During the alignment with the top management team, the CEO should appoint a couple of individuals from his team to take charge of the organization-wide alignment, which would involve communicating the following to all and sundry: why are we doing what we are doing. These leaders take ownership to align with representatives who are typically influencers, from middle management, front liners, shop floor personnel, and even union members. Next, this core team is formed with representatives from top management, middle management, and front liners; this team takes charge of rolling it out across the firm. This is achieved through creation of small leadership alignment teams (LATs) comprising business and functional leaders and the Lean Change Agent, who travel across the globe to communicate to all employees on the six questions. Communicating to all employees "why we are doing what we are doing" is a prerequisite that should be effectively done before starting the Lean transformation.

29

Why a Common Understanding of Service Is a Must in Lean for Service

Whether it is a manufacturing company or a service company, "service" is an integral part of customer delivery. But, if you ask a cross section of employees what service means to them, you will find that they come up with different responses or may even just give you a blank look. This is ironic, but in a service organization, employees not knowing what the word *service* means to them indicates the need to reevaluate the service strategy of the organization.

From my experience, I have seen that this is a problem plaguing many service companies. When customers are not able to verbalize what service means to them, it is just a symptom of a deep-rooted malady facing the company. It is a manifestation of leadership not taking service as a key business proposition. You may be wondering how Lean is related to this. We do not realize it, but great many resources are wasted because of such a fundamental gap in definition.

Actually, it is simple: When your employees do not know what service stands for, how do you expect them to provide consistent service? This not only has an impact on customer satisfaction but also confuses the customers regarding where the organization stands as far as service is concerned. A large amount of employee bandwidth is lost on something that could have delivered better customer engagement. This is a huge organizational waste that could have been avoided by better strategic alignment. However, if you do a root cause analysis, you will find, among others, the following key reasons for these delivery dissonances:

- Lack of a well-defined service intent that defines what constitutes a service experience
- A gap between the brand personality and customer experience

- Customers not engaged in what to expect from the company
- Employees not trained on what should be expected of them

So, how does one arrest this problem? This begins with the establishment of strategic business objectives, which is an integral part of the tetra-headed outcomes of Lean.

The deployment of the service strategy begins with the creation of the service intent. The service intent begins with what service means to the company and how this would be penetrated deep and wide within the company and even to the customer.

TABLE 29.1

Approach to Service Intent Deployment

Step	Details
Delineate service intent	This details how the organization would like to treat its customers. What should a customer expect when experiencing the company, its employees, or products or services? The service intent should clearly state what the customer can expect from the company and also tell the employees what is expected of them to meet customer expectations.
Reposition brand values	The company develops a brand image to embed service dimensions of customer delivery. The brand should reflect the service personality that the company stands for.
Communicate to employees	The company communicates to all its employees the service promise that it has created for the customer. It also seeks inputs from the employees on the proposed service promise. Employee involvement is the key to making them feel that they are a part of the development process.
Redefine roles	Ensure that all employees dealing directly or indirectly with customers spend adequate time to meet the service promise.
Create a service standard	Build a solid standard for all points at which the customer comes in contact with the company. These are activities required to meet the customer requirements. The endeavor should be to install an engaging experience at each stage of the customer experience. These standards help the service intent to become a reality.
Train, train, and train	Deliver role-specific training programs to employees. Focus on programs that instill service aptitude and develop people skills. Run regular refresher programs to keep the team updated. There should be an equal mix of technical and soft skills in the training programs.
Implement	Implement processes that will work toward supporting and meeting the service intent.
Measure	Track the progress and measure the experience of customers with the company. This could be done through various metrics and audits.

The strategic service intent states the experience that a customer can expect of a firm. This should resonate in every interaction that the organization has with the customer. The strategic service intent is driven by the brand values that the organization defines for itself. All employees who interact with customers have to demonstrate the brand's personality through customer experience. The brand is a promise made to the customer and aggregates as a collection of perceptions in the customer's mind. It is the brand that shapes the expectations of the customers. So, it is imperative that there should be consistency between what the brand stands for and what is delivered by employees who come in contact with customers.

Table 29.1 shows an approach that companies can follow to instill the same understanding of service among all employees.

There is one thing that should be kept in mind: Superlative service delivery requires committed and engaged staff who are kept motivated with a culture that rewards great service performance. Last, but not the least, top management has to treat service as a strategic imperative and competitive differentiator and not just indulge in lip service.

30

Who Are the Custodians of Your Process?

In a manufacturing shop floor, the processes generally have well-defined owners. These individuals take complete ownership of the deliverables of process outputs and manage all the issues pertaining to them.

However, this is often not the case in service organizations. Lack of process ownership results in a large number of inefficiencies and affects the overall effectiveness of the process. This is often because of the functional structure of the organizations. While processes flow horizontally, the functions work toward vertical agglomeration. This results in the following:

- *Local optimization:* The focus is on meeting the goals within the functions instead of the process goals. This leads to stagnant performance within each of the functional silos without really affecting the other silos.
- *Customers as orphans:* Nobody owns the customer and the customer's requirement. Each of the functions focuses on meeting their vertical objectives and the needs of the customer suffer.
- *Handoff misses:* This refers to misses that happen as the individual silos complete their process activity and pass on the baton to the next silo in the process loop.
- *Myopic measurements:* The metrics that become tracked within each of the silos do not give a complete overview of total process performance and how effectively it meets the customer requirements.
- *Missed process outcomes:* The outcome of the total process is not owned by anyone. This results in processes often not meeting their desired results.
- *Process dissonance:* There is no one to manage the coordination of different processes.

If the company is cast around processes and value streams, driving process ownership is easy. However, it requires structural change in organizations built around functions. This may not always be easy. Carrying out a structural change could actually derail the entire thought as the company may not really be ready to undergo the pain.

This is where you need to find out hybrid approaches that an organization can follow to drive process ownership.

- Create a team comprising senior members from each of the functional silos who jointly own the entire process (Figure 30.1). This team should be made responsible for the overall performance of the process. Members meet on a regular basis to manage the issues pertaining to cross-functional coordination.
- Make leaders from the top management team responsible for the core processes of the company. These individuals could be one or two levels below the chief executive officer (CEO). They should essentially be in a position to drive actions of functional leaders toward achievement of larger objectives of the process. This role could be played in rotation, but it is important that ownership remains for at least a period of 2 years (see Figure 30.2).

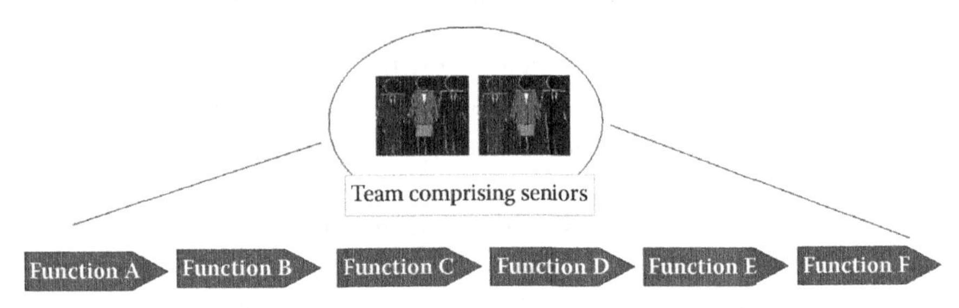

FIGURE 30.1
Joint ownership of process functions by senior members.

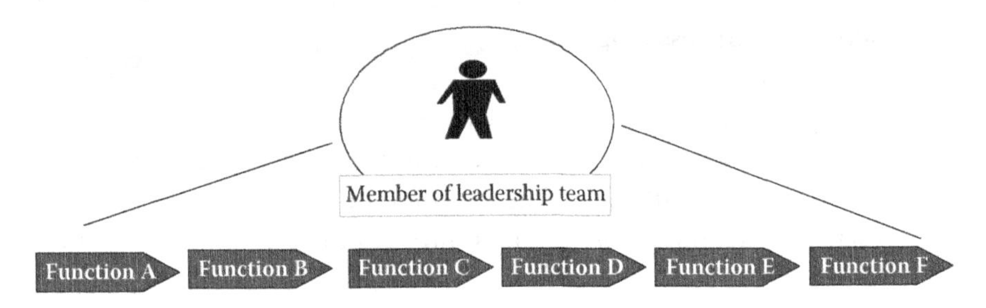

FIGURE 30.2
Members from the leadership team own the process.

TABLE 30.1

Roles of a Process Owner

1. Takes ownership for overall performance of a process
2. Owns the customer of the process
3. Takes responsibility for managing change across the process
4. Manages interface, handoff, and cross-functional issues
5. Shares process metrics and monitors performance
6. Is responsible for driving improvements across the process
7. Ensures process change happens and ascertains the impact of the change of functional and geographical capabilities
8. Identifies best practices that will improve the overall process effectiveness
9. Establishes strategies and priorities for investments in the processes
10. Helps to integrate people, technology, and leadership in a process

Driving ownership of processes that cut across geographies can be arduous. It requires solid leadership and commitment to manage coordination and handoff issues pertaining to processes. In such a scenario, the process owner has a role to develop globally aligned processes that address region-specific needs. Table 30.1 summarizes the key roles of a process owner.

31

Just Not Larger Projects

Implementation of Lean requires creation of an engine that allows improvement projects of all complexions to be completed. I have observed organizations embarking on a Lean journey doing a few improvement projects and thinking that the right thing is being done. Another approach that organizations often take is to focus only on high-impact projects. For them, the focus of the Lean transformation is only to do high-impact projects that have an impact on costs and revenues. While there is nothing wrong with such projects, the approach is incomplete. What is recommended here is that we follow the large improvement project-small improvement project (LIP-SIP) construct of improvements.

So, what is the LIP-SIP construct of improvements? It is an approach to improvements wherein LIPs are backed with a large number of SIPs (Figure 31.1). The following describes LIPs and SIPs in more detail:

Large improvement projects: These are large cross-functional projects taken up on key processes to deliver benefits to the bottom line. These projects take up to three to four months to complete and are centered on experts of Lean. These projects are driven by senior leaders and require the participation of a team representing all stakeholders, such as those involved with process, product, policy, risk, information technology, and so on. These projects have an objective to bring about a sea change in the way processes currently perform in the company. These projects are either sponsored by the top management or are led by middle management.

Small improvement projects: These are short-duration projects to be completed within two to three weeks that are taken up by associates working on the shop floor using elementary quality improvement tools. The target of these projects is to address local issues, which do

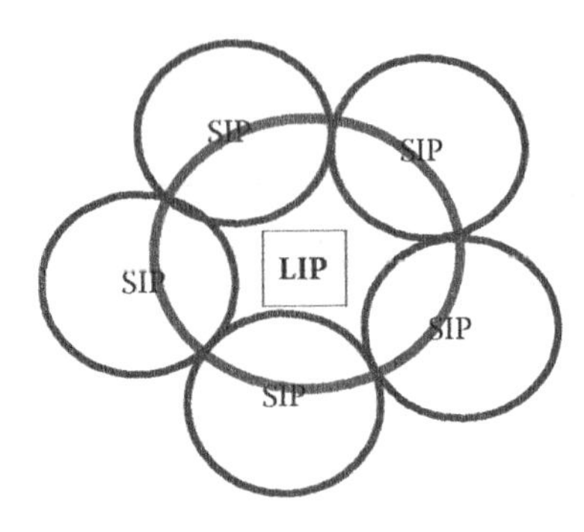

FIGURE 31.1
LIP-SIP construct of improvements.

not get on the radar of the leaders driving large projects. These projects have a major objective to garner involvement of the employees in improvements and build a culture of improvements *by the employees, for the employees, and of the employees*. The local leaders who could be a part of the junior management team sponsor these projects.

Table 31.1 gives examples of projects that fall in the categories of large projects and small projects.

Both LIPs and SIPs are endowed with some inherent strengths these complement each other and are a must for building a solid improvement fabric for the company. Large projects are performed on critical business processes of the organization, while small projects target local workplace issues and small work processes and procedures. The beauty of small projects is that they do not require large investments and help to build a solid foundation of quality by rooting a culture of continual improvement.

TABLE 31.1

Examples of Large and Small Improvement Projects

Serial Number	Type of Project	Sponsor	Example
1	Large improvement projects	Top management	Improving the way mortgage business is done by moving from a brick-and-mortar model to leveraging multiple channels
2		Middle management	Reduction of end-to-end disbursement lead times of mortgage processing
3	Small improvement projects	Junior management	Improving the productivity of processing associates in a credit shop

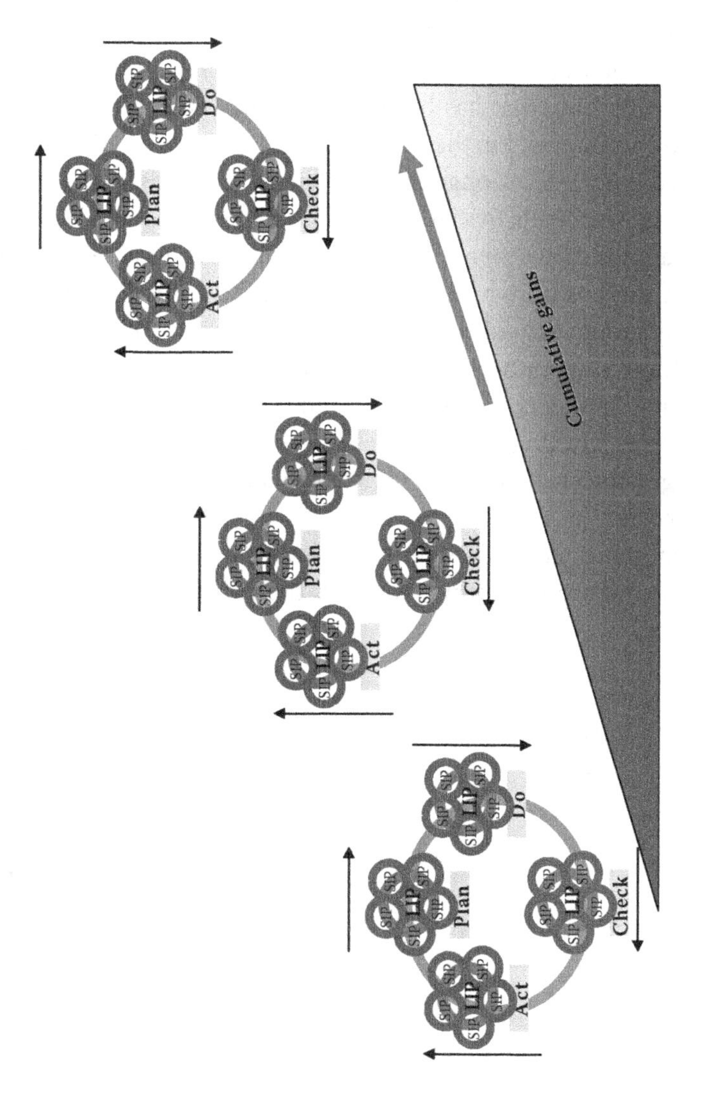

FIGURE 31.2
Small and large improvement projects in the Deming wheel of the company.

The employees take them up voluntarily to address specific issues that have an impact on their workplace for their benefit. When organizations just focus on large projects, improvements are centered on a set of experts who arrive, perform, and carry out improvements.

In the eyes of the shop floor individuals, large project implementation is about people like Lean Mavens, swooping into their workplace like extraterrestrials, facilitating a set of solutions, and then flying out. To them, it is adopting something that has been triggered by someone else and that they have to internalize. Further, it could appear to be an improvement that has been enforced for ongoing maintenance and sustentation. And, remember, improvements that are pushed by someone else and have been adopted in a halfhearted manner cannot sustain and gradually deteriorate. Even if there is a high degree of acceptance of such solutions, the chance of any system degrading over a period of time are imminent. Given this context, it is imperative that all large projects are backed with a large number of small projects, which are led and taken up for implementation by the associates working in the processes on the shop floor. This implies that one large project should be associated or followed by a large number of small projects.

For the larger organization, this would mean a Deming wheel of improvement (Figure 31.2) functioning in the company embedded with LIPs and SIPs.

32

White Spaces: A Great Lean Opportunity

Geary A. Rummler and Alan P. Brache introduced the concept of white spaces in 1991. White spaces are the areas between the boxes in an organization chart or the area between different functions that do not have ownership. According to Rummler and Brache, the opportunities for improvement are maximum in these places as they do not have any ownership, and efforts taken fall between the cracks. This concept is relevant in the implementation of Lean in an organization.

From a Lean perspective, each of these spaces is a hatchery of wasteful activities and provides a great opportunity for operational excellence enhancement. To ascertain the opportunity for Lean in an organization, it is a good idea to list all the white spaces in the company is. The larger the number of white spaces in the company is, the greater the opportunity for Lean will be. Each of these is a haven for hidden factories, which are on nobody's radar. In our discussions, we will also look at these issues. The following are a few things that could be symptomatic of potential white spaces in the company:

- Issues having no ownership
- Orphan infrastructure, equipment, capital items
- Headless function
- Locations, geographies, branches without a leader
- Time spent in coordinating

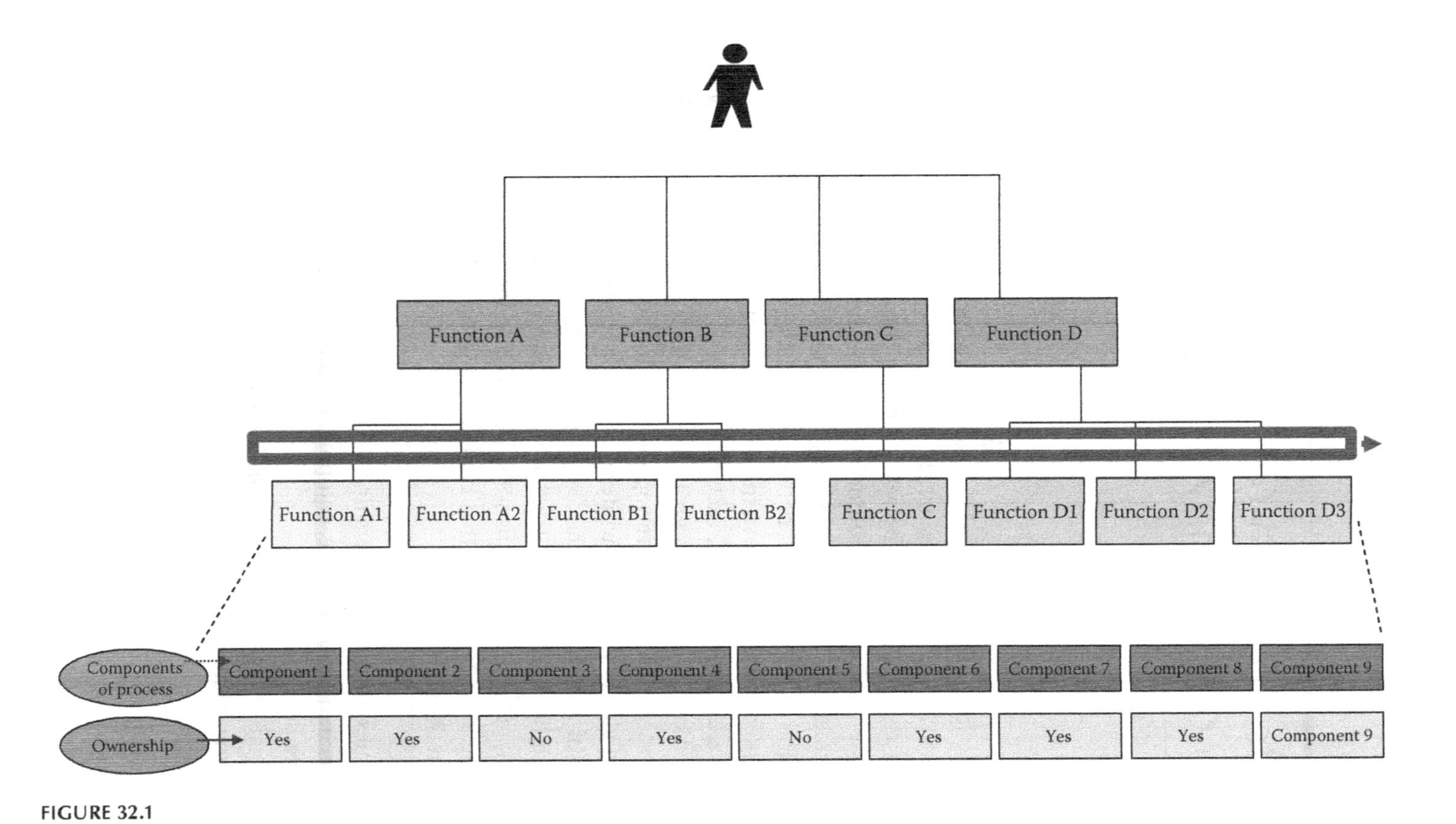

FIGURE 32.1
Finding white spaces in processes.

WHITE SPACES IN PROCESSES

A handy approach to finding white spaces in a process is to break a core process into broad components and then look for their owners (Figure 32.1). Components without ownership provide a clear opportunity for action. Installing process ownership is probably the first step toward managing these white spaces.

STRATEGIC OWNERSHIP MATRIX

Given the issues of ownership, it helps to create an inventory of all the key issues affecting the organization. So, how does one go about doing this? The following approach can help:

Step 1: List the Strategic Business Objectives of the organization (for a template, see Table 32.1).

Step 2: Once the Strategic Business Objectives have been listed, the top management team should brainstorm the variables that have an impact on these objectives. An approach as shown in Figure 32.2 could be used for this purpose.

Step 3: Having come up with strategic variables, these should be listed as in Table 32.2, with clearly defined ownership.

Although this matrix may not drive ownership of all issues in the organization, it can at least ensure taking care of the issues affecting the Strategic Business Objectives.

TABLE 32.1

Template for Listing Strategic Business Objectives

Strategic Business Objectives

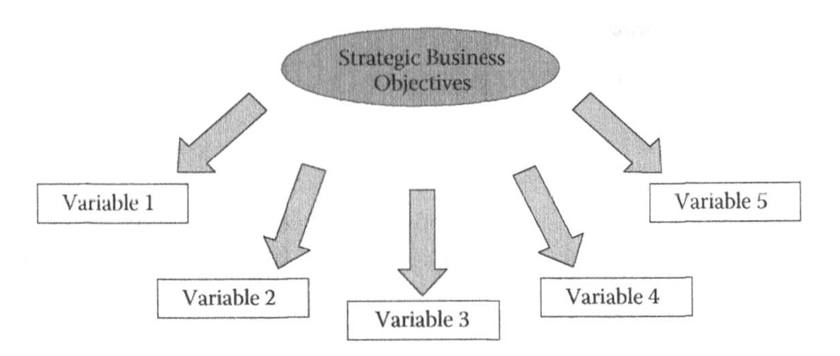

FIGURE 32.2
Key variables affecting the Strategic Business Objectives.

TABLE 32.2

Template for Strategic Ownership Matrix

Strategic Business Objectives (SBOs)	Key Variables Impacting the SBO	Owner	Stand-In

33

Does Your Organization Have a Standard Approach to Solve Problems?

Leaders and managers are hired essentially to solve business problems. However, companies do not invest sufficiently to develop this skill. A lot of money is spent on competency development programs, but little is invested on equipping teams with problem-solving skills.

A key element to build a Lean culture is to make everyone in the company practice PDCA. So, what is PDCA? It stands for Plan-Do-Check-Act. The PDCA cycle was originally conceived by Walter Shewhart and later adopted by W. Edwards Deming, who is often referred to as the father of modern quality management. Deming referred to it as the Shewhart cycle, but the PDCA cycle is often referred to as the Deming wheel or Deming cycle (Figure 33.1).

The components of the PDCA cycle are shown in Table 33.1. The universal nature of the PDCA cycle is such that it can be applied to anything or any type of problem irrespective of the industry, company, or context. It can be used for any change process. To drive continuous improvement, the PDCA process has to be repeated again and again. To see how PDCA thinking superimposes with problem solving, let us refer to Table 33.2.

Organizations typically have a large number of problems to solve, but with limited resources. So, it becomes imperative that not only high-priority problems are selected that have an impact on business outcomes but also a structured approach is used to obtain the desired results. This is where a generic problem-solving process is required to institutionalize across the organization so that it can be used by all levels in different contexts.

Educating teams on this problem-solving process cannot be confined only to middle management and front liners; it also has to be used by

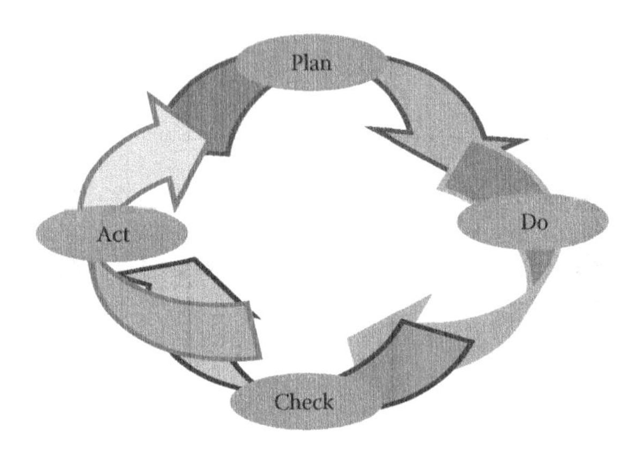

FIGURE 33.1
The PDCA cycle.

TABLE 33.1
Details of PDCA

P	Plan	Plan a change process.
D	Do	Carry out the required change on a small scale.
C	Check	Review the results: What went right? What went wrong? What was learned?
A	Act	Take action based on the results. If the results are successful, scale up the change. In case it did not work, begin the cycle again.

TABLE 33.2
PDCA and Problem Solving

P	Plan	Identify the problem. Define it clearly. Collect relevant data. Predict the expected change from the project. Brainstorm all possible root causes. Propose solutions. Prepare a plan for implementation.
D	Do	Go ahead and implement the solutions. The objective is to test the hypothesis of the solutions.
C	Check	Gather performance data on the solutions that have been implemented. Capture what was learned. Investigate if the outcomes were not as anticipated.
A	Act	Roll out the solutions across the entire organization or business system.

top management. Figure 33.2 is a generic problem-solving process used by Toyota. Organizations can also adopt problem-solving methods such as DMAIC (Define, Measure, Analyze, Improve, Control). A generic problem-solving approach helps teams emerge from firefighting to a structured approach to address workplace issues. Jeffrey Liker and Michael Hoseus, in their book *Toyota Culture—The Heart and Soul of the Toyota Way* (see Figure 33.2), talked about Toyota's problem-solving method (Toyota Business Practices, launched in 2005 in the United States).

Instilling a problem-solving culture results in teams becoming involved in improving their workplace and contributing toward business betterment. This can be satisfying for the employees, as they feel good about being able to contribute to the issues that their leaders want to solve. Organizations need to shun their focus on tools and really get into creating capabilities that allow employees to solve their day-to-day problems. All employees in the company should be taught the art of problem solving.

Every new entrant in the company, across the hierarchy, also needs to undergo this training. Leaders, while reviewing outcomes of projects, should look not only at the results but also at the process followed. Rather, if someone comes up with a result without following the proper process, the result should be treated as a blindfolded walk.

As a matter of fact, to sustain a Lean fabric, daily problem solving should be a way of life. Given the universal nature of the problem-solving process, it can be used for a wide range of projects in different contexts

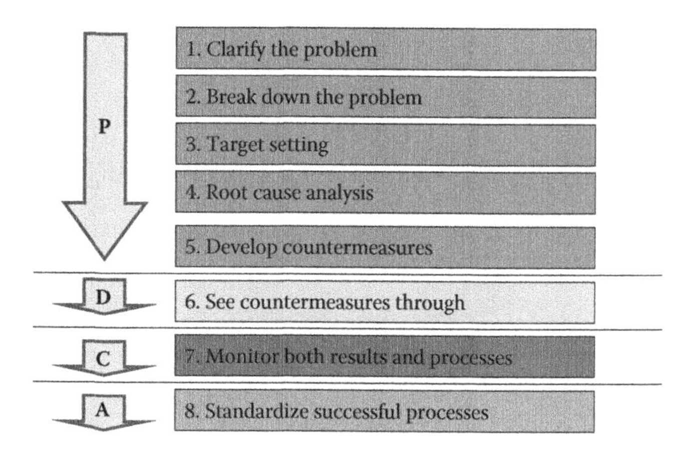

FIGURE 33.2
Toyota Business Practices (problem-solving process at Toyota). (Adapted from *Toyota Culture—The Heart and Soul of the Toyota Way*, Jeffrey K. Liker and Michael Hoseus, New York: McGraw-Hill, 2008.)

and functions. The endeavor should be to shift the culture of the company from that of a blame game to that of problem resolution and prevention. Inculcate a culture wherein problem solving not only is the job of the Lean office but also involves everyone in the company. Of course, the type of problems taken up by top management will be quite different from the ones taken up by front liners. The former will have a strategic bent, while the latter will be tactical in nature. However, the approach followed will be the same.

I would recommend that the problem-solving process be strengthened by following the A3 process, discussed in the next lesson.

34

Shun Verbosity and Long Presentations: Adopt A3 Thinking

A3 thinking is the practice of using a single sheet of paper for solving problems, sharing proposals, and strategic planning. Because paper 11 × 17 inches (A3) is used for this purpose, it is called A3 thinking. Developed in the Toyota stable, the concept can be used in any type of industry and forces teams to share their story in a short and standardized manner.

The A3 sheet can be used for problem solving, strategic planning, proposal writing, and status reporting. The focus of A3 in this chapter is on problem solving.

The structure of the A3 sheet is such that it provides a brilliant approach to PDCA (Plan–Do–Check–Act) management. PDCA is Deming's approach to problem solving (see Chapter 33). The A3 report actually provides an approach to problem solving and creates a story line of the way a problem has been resolved. Figure 34.1 shows the template for problem solving. The sequence of problem solving should be according to the numbering of the steps (see Figure 34.1). The seven steps of the A3 report follow a PDCA approach; this is summarized in Table 34.1. As you can see, steps 1, 2, 3, and 4 comprise 50% of the A3 sheet. This shows the importance of planning in problem solving. An important thing to learn here is that we need to spend a lot of time understating the current state before getting into implementation. It also acts as a visual tool wherein the problem details are scripted and the members become involved in giving inputs to solve it.

Use a pencil to write the details so that things can be written and changed based on inputs received from teams. The steps can be changed by the companies based on their context, but the general approach to be followed is the same. The focus should be to make the problems visual so that they can be understood easily.

Name of the company: ______, Business unit: ________, Workplace: ______________, Project name: __________________, Project ID: __________, Team leader: _____, Team members: ______________	
1. Background	**5. Countermeasures**
2. Current state	
3. Goal	**6. Confirmation of results**
4. Root cause analysis	**7. Follow-up actions and sustainability imperatives**

FIGURE 34.1
Template for A3 problem solving.

TABLE 34.1

PDCA and Problem-Solving Steps

PDCA	Steps
Plan	Background, current state, goal, root cause analysis
Do	Countermeasures
Check	Confirmation of results
Act	Follow-up actions

However, how do you implement A3 thinking in your company? The following are a few things that you can do to accomplish this:

- Take a strategic call to adopt A3 sheets.
- Train all employees on A3 problem solving.
- Mandate teams and individuals across hierarchies to use A3 sheets.
- Put a ban on long presentations and make A3 sheets the only way to solve problems.
- Managers and leaders need to be adept in interpreting the finer nuances of A3 problem solving.

Like problem solving, the A3 sheet can also be used for strategic planning (Figure 34.2) and proposals (Figure 34.3).

Just imagine what A3 thinking can do for an organization:

- Eliminates long problem-solving presentations.
- Reduces number of hours spent on making long presentations.
- Forces teams to understand the problem statement adequately before proceeding to a solution.
- Brings visibility to the problem-solving process.
- Brings simplicity.
- Facilitates systematic thinking among employees.
- Helps seniors mentor their team members: any change in a process requires the buy-in of the bosses, which is a great mentoring opportunity.
- Avoids a communication gap between teams as the problem-solving approach is before them and they exactly know the flow.
- Prevents teams from taking shortcuts to problem solving.
- Acts as a knowledge repository.
- Results in better and fewer meetings.

1. Last year's performance

2. Reflection of last year's performance

3. Current year's objectives

4. Justification for this year's focus area

5. Detailed action plan

6. Critical success factors

7. Before and after metrics

8. Issues pending to be resolved:

FIGURE 34.2
A3 template for strategic planning.

- Eliminates long project review meetings as stand-in reviews would happen at the workplace or where the action happens.
- Compels a standardized approach to problem solving using the PDCA cycle.
- Fosters better teamwork and collaboration among team members.

To know more about A3 thinking, I recommend consulting the research work of Durward K. Sobek II and Art Smalley. They have written an excellent book titled: *Understanding A3 Thinking* (CRC Press, 2008).

Name of the company: ______, Business unit: _______, Workplace: _____________, Project name: _________________,
Project ID: _________, Team leader: ____, Team members: _____________

1. Background	**5. Plan highlights**
2. Current state	
3. Proposal details	**6. RACI* and time lines**
4. Dissection and analysis	* Responsibility, Authority, Consultation, and Informed

FIGURE 34.3
A3 template for proposal writing.

35

What Metrics Should You Have?

In the manufacturing sector, it is common to measure the performance of processes on a continual basis. The processes are visible, and measuring them comes naturally. Also, the improvement journey of manufacturing processes has been under continual scrutiny for more than 50 years.

The story is different in the service sector. The journey of improvements in the service sector is very old, and relevant processes may not always have the required metrics. The process in the service sector is often not visible, which makes it difficult to even appreciate that an activity is a process. This is especially true for functions that are traditionally not thought to be driven by processes, such as sales, marketing, and finance.

However, things are changing in organizations. Service sector companies have begun installing metrics. But, when scrutinized closely, one often finds that metrics are not relevant and correct. The problem is that the focus is often on installing metrics and not on the benefit they will deliver. So, metrics are installed, they are measured, man-hours are spent to analyze them, but they do not deliver much value. Thus, despite the intentions being pious, organizations inadvertently create waste, which could have been avoided if they had looked at installing the right metrics. This is what I call the waste-of-measurement system.

So, what are the pointers that companies should remember before installing metrics? I call them the A-to-H principles of installing relevant and waste-free metrics in a company on a journey toward Lean transformation.

Table 35.1 summarizes Debashis' or Deb's A-to-H principles of installing the right metrics in an organization.

I would recommend that Lean change leaders, business leaders, process owners, and value-stream owners ascertain the overall effectiveness of the measurement system. This can be done by following the template shown

TABLE 35.1

Deb's A-to-H Principles of Metrics

Attribute	What Does It Mean?
A Action Oriented	The metrics should be presented on dashboards in such a manner that the process associates can comprehend them easily and take action. This can be achieved by following the traffic light colors of red, yellow, and green to represent performance.
B Big Picture Connect	The metrics should be aligned with strategic business objectives. This should include meeting the expectations of shareholders, partners, customers, regulators, employees, and society.
C Customer Led	The metrics should meet customer requirements and should have linkages with operational levers.
D Digitized	The collection of metrics should be digitized. Avoid manual data collection processes. Even if it takes time, gradually the organization should move toward automated data collection, which later is converted to a digitized dashboard.
E Employee	Whatever metrics are decided, it is imperative that they should have the involvement of the employees. It is OK to use an outside resource, consultant, or member of the Lean office as a facilitator, but never push down the metrics without the buy-in of the employees.
F Familiar	The metrics should be standardized across business units of an organization. There should be a common currency that helps everyone understand the process performance irrespective of the functions and levels.
G Guiding	The metrics should include leading and causal indicators that help predict the outcomes of the process and allow action to be taken before something goes wrong.
H Holistic	The metrics installed should include end-to-end metrics cutting across silos, which gives a holistic approach to measure the way the process performs.

in Table 35.2. The blank spaces can be populated by using the scoring pattern given next.

SCORING PATTERN

The following pattern can be used for scoring:

What Is It?	Score
Does not exist	1
Partially exists	3
Fully exists	7

TABLE 35.2

Template for Ascertaining Effectiveness of Measurement System

Attribute	Business Unit/ Value Stream 1	Business Unit/ Value Stream 2	Business Unit/ Value Stream 3	Business Unit/ Value Stream 4	Business Unit/ Value Stream 5
A Action Oriented					
B Big Picture Connect					
C Customer Led					
D Digitized					
E Employee					
F Familiar					
G Guiding					
H Holistic					

INTERPRETATION

Table 35.3 could be used to interpret the scores. Remember, this is only directional and essentially tells you how far you are from an effective measurement system.

Having installed metrics, it is futile for organizations to stay with them forever. Typically, what happens is that once metrics are installed, no one questions them. All that happens is that newer metrics are added to the existing ones, and the list keeps on increasing. This especially happens after an improvement project is completed. As a result, some newer metrics are added, among other things, and the processes are inundated with metrics; prioritizing the important ones becomes difficult. Added to this is the fear that if metrics are removed, the performance of the process would be affected.

A good Lean organization continually questions the relevance of metrics to the business. Remember that metrics are just an enabler of business improvement; they cannot become the business of an organization.

Install a system in which the metrics are reviewed on a regular basis so that redundant metrics are done away with and newer ones are installed wherever required. I would recommend that all service organizations refresh their metrics once a year.

TABLE 35.3

Interpretation of Scores for Effective Metrics Management

Score	Interpretation
More than 49	The measurement system is effective and right for creating a Lean organization.
36–48	The measurement system is partially effective.
22–35	The measurement system is just being built, and a lot needs to be done.
Less than 21	The measurement system is pedestrian and not effective.

36

Is Employee Attrition a Problem in Your Company?

One of the biggest hidden inefficiencies in companies is when individuals are recruited but do not stay for long. It leads not only to heavy rework but also a lot of resource wastage. Monies are spent on training or inducting people only to lose them later. To my astonishment, some leaders say that attrition is a reality, and we need to work a business around it. They rationalize this problem by making the following statements:

- Attrition is an industry-wide phenomenon, and we have to live with it.
- Attrition is a problem of an emerging market wherein the growth of the economy has created plenty of job opportunities.
- Attrition has only been at the grassroot level—there has been no attrition in the leadership team.
- In today's context, you cannot expect a newcomer to stay in an organization more than a year or two.
- This is a Gen-Y problem.
- Beyond a point, I do not worry about attrition. I need to just ensure that my diamonds [top performers] stay in the organization.
- We have been a great training ground. Our colleagues get picked up by the competition.

All of these statements are manifestations of an "un-Lean" mindset. Instead of allowing the problems to surface, these individuals are accepting the situation to be something that should not cause worry. Further, they go ahead and actually *budget* these attritions—a typical "just-in-case" mindset.

Have you ever wondered why there is employee attrition? More often than not, it is because the recruitment process does not receive its due importance. It is treated as just another of those human resources processes. It is defined but not to the detail required. It is defined but only partially and not adhered to the fullest. Recruitment is not treated as a strategic function in the company.

It shocked me when a business leader, during a discussion of Lean implementation, mentioned that he needed a few "working hands" to make Lean a reality in his business. I was amazed at this mindset of treating employees as working hands or robots who will just take instructions and execute them.

Getting the right team set is probably the first step toward creating a Lean organization. Remember, motivated employees will not only provide great service to customers but also create a great Lean-thinking organization.

Many times, the right people do come on board but leave because of the environment provided by the organization. They may not be fully engaged in the endeavors of the company. There could be concerns about the manner in which the leaders treat their employees. The employees may not have a sense of belonging toward the firm, and they may be feeling devalued. They may not be used to their full potential and unable to see where their career is moving. They may have stopped learning and adding value to themselves. They may not have leaders around them to whom they can look up.

So, what are the things an organization should do to arrest attrition and ensure that the right individuals come on board?

- Be clear on the type of individuals who you would like to hire in terms of both technical and adaptive skills.
- Focus more on attitude than technical skills. (I have seen companies implementing Lean focusing more on whether the individual knows relevant tools and techniques. Remember, you can always train them on Lean tools, but if you do not get the right mindset, the game is lost.)
- Ensure that the candidates have competencies that meet both the short-term and long-term goals. This means that the candidates should not only meet the current requirements but also have the skills or potential to grow into future roles.
- During the interview, unfurl the behaviors that may have been demonstrated in the past.

- The candidate should manifest team-adding skills or traits.
- Install a rigorous hiring process—despite all the pressures, never take shortcuts.
- Relentlessly communicate to the outside world on the type of talent you are seeking.
- Only hire those individuals who meet the desired values of the firm.
- Never paint a rosy picture during the interview process; clearly communicate the opportunities, expectations, and challenges of the job.
- Look for candidates who are high on problem-solving skills and have humility and respect for other people's work or contribution.
- All new inductees should go through a solid foundational training comprising not only the functional skills but also information about the company and its values, beliefs, and culture.
- Shun the thought that there is a talent shortage so you should hire whoever applies.
- Make the performance management system of supervisors accountable for retaining, nurturing, and developing talent.

My advice to companies is to draw value-stream maps for the entire recruitment process, which comprises attracting talent, the hiring process, the on-boarding process, and the first 90 days in the organization. This will clearly bring out the activities that do not add value and could be waste. Value addition in such a process would be the identification of all those activities that do not help in development and growth of the new recruit. Make this value-stream map for all key roles in the organization. Remember, the output of this value stream is to obtain the right quality and quantity of talent, timely recruitment, and induction of new recruits in time.

37

Inventory in a Services Organization Can Be of Various Hues

One of the key objectives of Lean is inventory reduction. This can be quite a challenge in a service context that includes financial services. In a manufacturing company, inventory is tangible and includes things such as raw materials, semifinished goods, rejects, and so on. These are visible, so taking actions on them becomes easy. The objective is that reducing inventory will unfurl wastes. Having learned to see wastes, attacking them becomes easier.

However, in a service context, this can be difficult. Inventory may not always be tangible and visible. So, working on its reduction can be a challenge.

The typical inventory in a service industry could be the following:

- E-mails
- Files
- Documents
- Data
- Information
- Electronic reports and documents

To understand what an inventory in a process is, you need to use the "use" and "consume" principles of processes. In every process, there are things that are used and there are items that are consumed.

Consumed are those items that transform and change their shape, form, and state as a part of process completion. Let us take an example from the automobile world. Items that are consumed while manufacturing a car include things such as steel. The form and shape of the raw material change and transform into the final product. Many times in a services context, the items that are consumed may not have a physical transformation, but their physical state may change. In a mortgage finance business, the status of the form changes from an unapproved state to an approved state.

Used items are those required during processing but do not transform by the end of the process. These items are more like resources in a process. In the case of a car manufacturer, such items could be people, machines, and so on. In mortgage processing, the typical items that are used include people, computers, infrastructure, and so on.

So, when you have to identify inventory in service processes, ask a question to see if it is consumed. If the answer is affirmative, it is a target for inventory reduction. We will look at Little's law, which is handy while working on inventory reduction. We will understand more about Little's law in Chapter 54.

Making all employees understand and appreciate the inventories in service processes is the first step toward waste optimization. Just remember, humans who are involved in the processes can never be treated as inventory. These are the people who run the process. So, never treat customers as inventory. However, there are some exceptions. While working on Lean for a human resources process such as recruitment, humans are to be treated as inventory. Also, in customer-facing entities such as retail branches, stores, or supermarkets, when you are working toward reducing wait times, customers will be treated as inventory.

From the Trenches

Excess cash holding in bank branches calls for a Lean intervention. Bank branches holding more cash than required affects the profitability of the branches. This is because cash available but not deployed is a lost revenue opportunity for the bank. Branches holding large volumes of cash, apart from being a big risk, affect the profitability of the bank because cash kept in a safe does not generate any interest.

Under my guidance, a team took up a project to reduce the average cash holding of branches in a large financial service conglomerate. The principles of Lean were used to solve this problem. As part of the roll out, the following processes were used:

- Waste identification
- Cycle time study
- Small batch size
- Standard work procedures
- Flow
- Pull

TABLE 37.1

Wastes Observed in Cash Projects

What Was Observed	Type of "Ohno's" Waste
Cash bundling not done by tellers	Waste of underutilized people
Excess cash indented	Waste for overproduction
Cash vehicle movement	Waste of transportation
Cash waiting to be bundled	Waste of waiting
Leftover processed cash involves rebinding and packaging for offloading	Waste of overprocessing
Unprocessed cash lying in vault	Waste of inventory

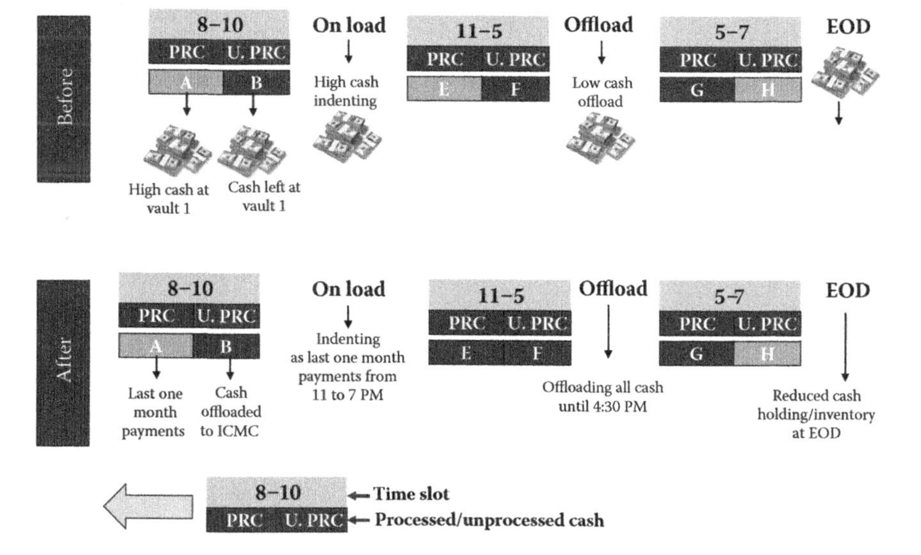

FIGURE 37.1
Pull in a cash management process in a retail bank branch. EOD, end of day; ICMC, integrated currency management chest; PRC, processed cash; UPRC, unprocessed cash.

As a part of project execution, the types of wastes shown in Table 37.1 were identified. Figure 37.1 shows how pull was created in the process.

The intervention saw around 40% reductions in cash holding.

The objective of the example is to elaborate that Lean principles can be applied in almost all sections of the bank.

38

The Functional Crevices Provide a Great Lean Opportunity

While Leaning your organization, one of the things that you should look at in each of the functions, departments, or roles is the waste. It has been observed that, over a period of time, there are redundancies that settle into these areas and need to be eliminated. These can be in the following form:

- *Efforts not valued:* Functions and departments are wasting time on activities that are not relevant.
- *Overlapping of roles:* Two or more roles are doing the same activities or activities are being duplicated.
- *Duplication of functions or departments:* Similar functions or departments are in other parts of the organization.

So, how does one go about Leaning an organizational function or structure while addressing these issues?

LEANING EFFORTS THAT DO NOT ADD VALUE

Step 1

Departmental Overview Listing

List all the vital details of a specific department or function. You can use the template shown in Table 38.1.

TABLE 38.1

Functional or Departmental Overview Listing

What is the mission of the department or function?
What are the key deliverables of the department or function?
How does the department or function contribute to the larger organizational objective?
What are the key activities of this role?
Who are the customers of the department or function?

Step 2

Value Contribution of Listed Activities in a Function or Department

The value contribution of the activities listed in the previous step needs to be ascertained with respect to the department or function. This is explained visually in Figure 38.1.

Value-Added Activity

For an activity to be value-added in a department's or function's perspective, we should look at it through a Lean lens and answer the questions in Table 38.2. For an activity to be value-added (in a department or function), you should receive affirmative answers to all the questions in Table 38.2.

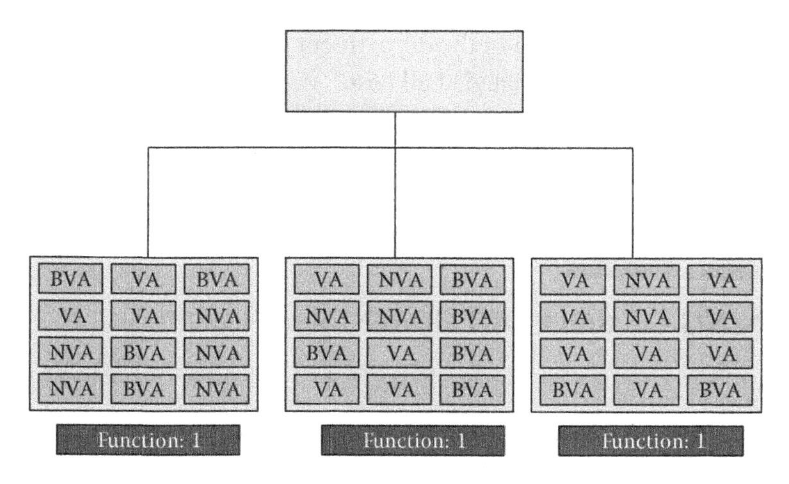

FIGURE 38.1
Value-added (VA), non-value-added (NVA), and business value-added (BVA) activity in function.

TABLE 38.2
Questions for Value-Added Activity in a Department

Serial Number	Questions	Yes/No
1	Is the activity being done for the customer?	
2	Does the activity contribute toward achieving the organizational objectives?	
3	Is this something that is being done for something that is "right first time"?	

Business Value-Added Activity

Beyond these activities, there could also be actions that do not meet organizational objectives. In such a case, the questions listed in Table 38.3 need to be answered. For an activity to be business value-added (in a department or function), you should receive affirmative answers to all the questions in Table 38.3.

TABLE 38.3
Questions for Business Value-Added Activity in a Department

Serial Number	Questions	Yes/No
1	Is the activity being done for the customer?	
2	Is the activity required for control or to meet statutory, regulatory, or environmental stipulations?	

All other activities done in the department are non–value added. These activities should be eliminated at all cost.

Remember, these activities within functions should not be mixed with the tasks that are scrutinized while looking for wastes in processes.

LEANING OF ROLE DUPLICATION

Step 1

List All the Activities Done in a Department or Function

Brainstorm all the activities done in a department or function comprehensively. The template given in Table 38.4 describes some of the key activities done by the department or function. Ensure that each of the activities has a unique number and nomenclature. This will help with unique identification and easy treatment of the data.

Step 2

Inventory All the Activities in the Template

The input of this can be obtained from the template in Table 38.4. Ensure that each of the activities done is listed and the corresponding time spent on each is mentioned. Look at the job description of each and ascertain their function and the time required to perform them. If a job description is not available, obtain an idea of the role from the functional head. This is not the best option, but until the job description is in place, this approach can be adopted. It is imperative to mention that the observations are indicated at the end of each role for subsequent analysis or action. It should be noted that, for managerial roles, exact time and motion cannot be presented. For managerial roles, one only has to do time assessment.

Table 38.5 is an example of a few roles in an internal consulting function of a bank that underwent a Lean process.

For each of the observations, get into the root causes and work toward eliminating them.

TABLE 38.4

Template for Activities Inventorization

Department/Function:			**Head:**			**Date:**		
Role: 1			**Role: 2**			**Role: 3**		
Objective:			**Objective:**			**Objective:**		
ID	**Activity Name**	**Time Spent, %**	**ID**	**Activity Name**	**Time Spent, %**	**ID**	**Activity Name**	**Time Spent, %**
Observations:			Observations:			Observations:		

TABLE 38.5

Template for Activities Inventorization (Filled)

Function: Internal Consulting			**Head: Deb Cooper**			**Date: June 27, 2008**		
Role: Engagement Manager			**Role: Capability Building Leader**			**Role: Practice Leader**		
Objective: Engage Business Units to Adopt Practices of Quality for Business Improvement			**Objective: Create Competent Resources within Business Units So That They Can Carry Out Improvement Without External Help**			**Objective: Develop Quality Practices for Leveraging for Business Improvement**		
ID	**Activity Name**	**Time Spent, %**	**ID**	**Activity Name**	**Time Spent, %**	**ID**	**Activity Name**	**Time Spent, %**
	Business engagement	30		Capability strategy	20		Practice development	30
	Manage project execution	30		Capability execution	25		Knowledge scanning	10
	Practice development	20		Business engagement	40		Manage project execution	30
	Follow up with back office	20		Follow up with back office	15		Capability execution	30
Observations: Question why time is being spent on follow-up. Time spent on practice development is high. Are there issues with the practice leader? Time spent on business engagement is a good amount less.			Observations: Time spent on capability execution is less. Why does the person need to spend time on business engagement? Follow-up should be reduced or eliminated.			Observations: Why is the individual spending time to manage project and capability execution? The time being spent on knowledge scan is much less. Work toward eliminating the role overlaps.		

LEANING DUPLICATE FUNCTIONS

Over a period of time, similar functions or departments are created within the organization. During one of my Lean engagements in a financial services business, I saw that, despite having a customer services call center for the organization, a bank branch had put in place a team that worked on customer queries. This was a clear example of a duplicate office: A call center had been created within a bank branch, and the center was not required. The reason for this was to cover for "just-in-case" scenarios.

This is a big problem, especially in large organizations, where over a period of time functions are duplicated, which only adds to costs and hides inefficiencies. To address such a problem, a template such as the one shown in Table 38.6 can be used.

Table 38.7 shows an example of an assessment for functional redundancies in an organization. As you can see, over a period of time, redundancies have set in.

As you see from Table 38.7, there is something clearly amiss. The ratio of human resources professionals per employee varies across geographies and does not follow a uniform trend. On investigating, dissonances were found not only in numbers but also in levels. This was a case of a bank that had grown inorganically and inherited varied practices from its lineage.

TABLE 38.6

Template for Inefficiency Assessment

Function Examined:			**Objective:**				
Business: 1			**Business: 2**		**Business: 3**		
Geo 1	**Geo 2**	**Geo 3**	**Geo 1**	**Geo 2**	**Geo 1**	**Geo 2**	**Geo 3**

Note: Geo: geographies.

TABLE 38.7

Assessment of Functional Redundancies

Function Examined: Human Resource Function			Objective: To Ascertain the Functional Redundancies in a Financial Services Organization				
Business: Retail Banking			**Business: Corporate Banking**		**Business: SME Banking**		
Asia	**Europe**	**Americas**	**Asia**	**Europe**	**Asia**	**Europe**	**Americas**
India (2,000/15,000)	United Kingdom (70/2,000)	Toronto (120/5,000)	India (20/500)	United Kingdom (5/50)	India (45/1,000)	United Kingdom (18/200)	Canada (5/200)
Singapore (50/1,000)	Germany (10/100)	Argentina (2/20)	Singapore (5/50)	Germany (5/20)			Brazil (5/100)
Hong Kong (20/550)	Nordics (5/100)	Peru (55/1,000)			Hong Kong (50/5,000)		
			It did not operate in corporate banking space in the Americas.				

Note: Please note that the figures in the numerator show the human resource employees and the ones in the denominator show the total number of employees.

39

MIS Reports, MIS Reports, and More MIS Reports

We all know that a management information system (MIS) is a mechanism for collecting, processing, storing, and disseminating data in the form of information to facilitate the functioning of the management and allow proper decision-making. However, does your company have problems of too many MIS reports or management reports? Are reports generated by every function and department? Are reports prepared, but no one looks at them? Are MIS reports generated at the drop of a hat or with every change in leadership? All these are symptoms of deep waste in an organization. Multiple MIS reports generate a large number of problems in the organization; not only do they elaborate waste but also they have knock-on effects, such as the following:

- A large amount of confusion on definitions or interpretations
- Wasted energies in defending or holding one's turfs or reports
- Man-hours wasted in preparing reports that have little value
- Information technology (IT) storage space wasted to store data that are of no relevance
- Questioned data sources and data integrity issues raised
- Faulty decision-making

So, how does one go about resolving this problem? This can be effectively done by following the approach listed in Table 39.1. Do a detailed dissection through Lean methodologies and populate this template.

As a part of Lean transformation, list all the MIS reports that exist in the company using a template (Table 39.2).

TABLE 39.1

MIS Report Rationalization: An Approach to Leaning an MIS System in a Company

Steps	Details
Inventoried	List all the MIS reports using the Template in Table 39.2.
Stop	Stop all MIS reports that are being generated by all and sundry. This requires a mandate from top management.
Define	Define the objectives and the requirements of an MIS report by all levels of the organization.
Own	Co-opt a team that will have the responsibility of all MIS requirements of the company.
Detail	Detail the nitty-gritty of the MIS reports regarding what they will and will not contain.
Implement	Implement the detailed new structure for MIS reporting.
Review	Install a regime of regular reviews of MIS reports and their relevance.

The main reason organizations are inundated with MIS reports is because there is no ownership for the same. They are not given due importance and are looked at as unimportant enablers in a company's success. An MIS report should always be strategically aligned and should be designed in such a manner that it is relevant to the user. Make a team responsible for all the MIS reports for the company. If there are too many owners, the chances of errors and inaccuracies are imminent. Stop all MIS reports that may be generated by individual workplace owners or business leaders. Ensure that all data used for MIS come from a single source. Both these issues have to be audited regularly by team members. However, let there be sufficient flexibility to create newer forms of reports without really tinkering with the data source.

Issues pertaining to MIS reports can generate a lot of inefficiencies in business, and they have to be avoided at all cost. I would recommend that companies have an IT system for MIS that can be used enterprise-wide. Put a ban on use of isolated spreadsheets (like Microsoft Excel) to generate reports. Let a senior leader in the company also take ownership of all MIS reports so that they are controlled and managed as required.

TABLE 39.2

Inventorization and Analysis of MIS Reports in an Organization

Business Unit	Function	Departments (Within Function)	No. of MIS Reports Produced	List of MIS Reports	Objective of Report	Target Audience of the Report	Who Produces the Report	Man-Hours Required	Areas Covered in the Report	Remarks

40

The Role of the Lean Team Should Change over Time

We discussed installing a Lean team or Lean office in Chapter 11. However, the role and composition of this team should change as the implementation matures in the company. After a few years of deployment, it would not be wise to expect the Lean office to function in the same way as it did in the early years of the Lean journey.

In the first few years of Lean implementation, the focus of the Lean office should be on awareness creation, getting organizational engagement, and taking ownership for initiating improvements in various parts of the organization. The focus should also be to build capabilities across the length and breadth of the firm and to build a solid book of knowledge and experience as relevant to the company. It is a relatively large team comprising a large number of Lean Mavens and engagement managers. Their major focus should be to get the right engagement, hold hands, and demonstrate to teams how to carry out improvements. Remember that in this phase, the Lean office is one central team that takes substantial ownership for deployment by facilitating Lean adoption by the business units or value streams.

However, as time passes, the complexion of the team should change. The size of the Lean office should reduce by half. The implementation of Lean should shift to respective value streams and business groups. The business units and value-stream owners take ownership for Lean implementation and leverage it for reaping business results. Then, the Lean office is taken aboard only when specific technical inputs are required. In this phase, the Lean office focuses on capability building, best-practice sharing, and tracking ongoing performance. The Lean Mavens who were earlier with the Lean office gradually move into business units and act as

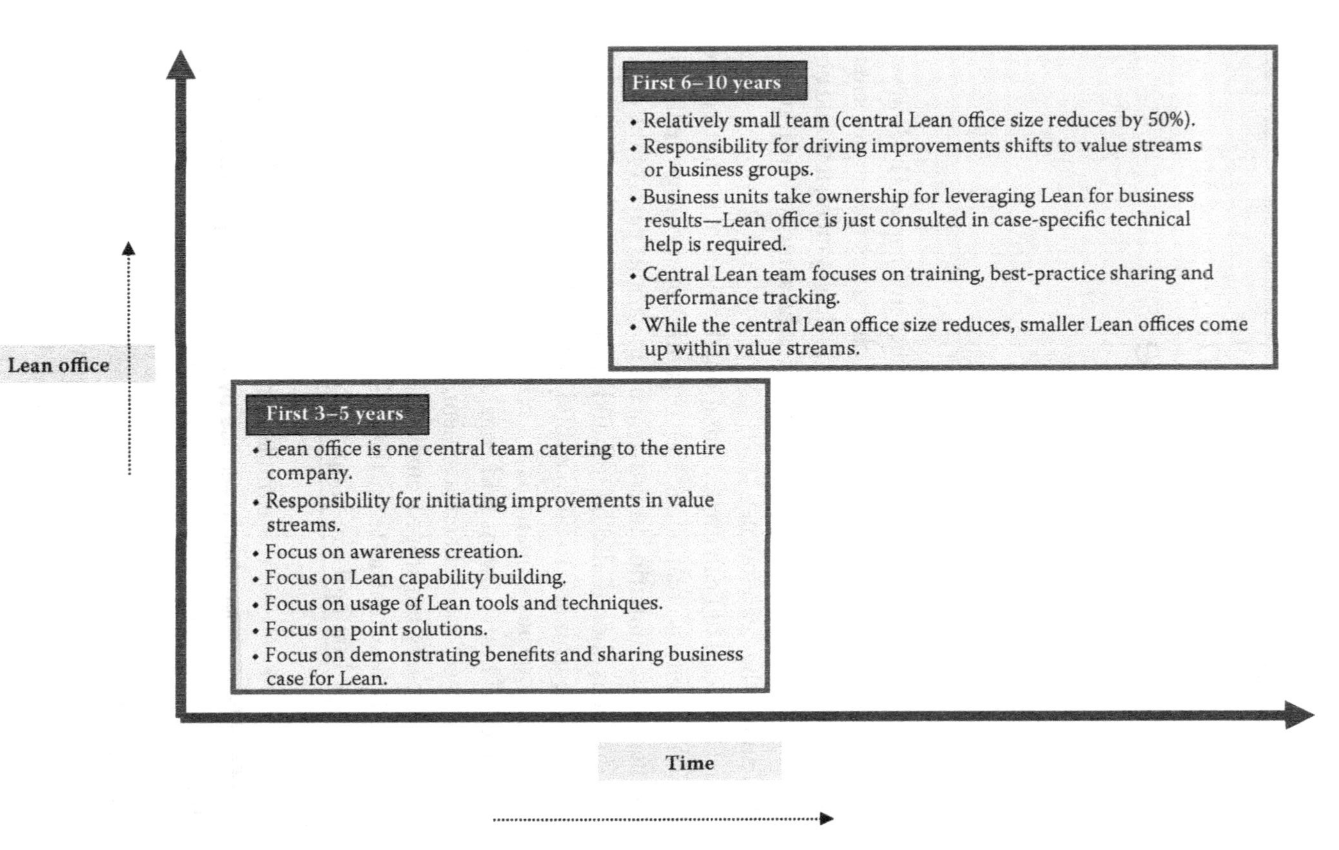

FIGURE 40.1
Changing complexion of the Lean office.

resources there. In this phase, the Lean office is also available for specific facilitation, such as project selection. As the Lean journey matures, the Lean office becomes decentralized, and each business group and value stream installs its own small Lean office. Remember that when Lean projects are done enterprise-wide, integrating the firm-wide gains can often be difficult, especially in large companies. This is because of lack of knowledge among Lean office teams of the value streams and customers these value streams cater to. Figure 40.1 summarizes the changing roles of the Lean office.

Table 40.1 summarizes the changing complexion of a central Lean office in a global financial services company having 30,000 employees and offices around the globe. The Lean Mavens move into the various value streams or business units, while the Lean Navigators take up lateral positions within businesses and value streams.

Remember that the role of the central Lean office has to change from being a demonstrator of "Lean as a value add" to that of being a "catalyst for business improvement." If this team is not able to reduce its size and ensure business units adopt the Lean principles for achieving business objectives, it has not done a good job.

To have decentralized Lean offices in the early days of implementation is not a good idea. This will not work as business and functional heads may not treat Lean foundation building as their goal when they are busy achieving business targets.

The biggest goal of the central Lean office during the first phase of Lean implementation is to ensure that businesses and value streams start leveraging Lean for their financial results.

TABLE 40.1

Changing Complexion of the Central Lean Office

Composition of Central Lean Office	**Phase 1 (3–5 Years)**	**Phase 2 (Beyond 3–5 Years)**
Lean Mavens		
Lean Navigator		
Lean Infrastructure Leader		
Lean Capability Leader		
Lean Change Agent		
Office assistant		
Total	**25**	**9**

41

Make Customers Service Themselves

One of the objectives of a Lean intervention should be to create an environment in which customers can serve themselves. Do not be amazed at what I am telling you. Some people believe that by making customers serve themselves, you will be alienating them. This is not true. This is an old shibboleth that one cannot cling to in today's fast-paced world of customer service. There are ample examples today to prove that self-service is something that the customers readily accept and find convenient.

So, what are the benefits of self-service? The following is a list of key benefits that an organization can reap from customer self-service:

- It provides better customer convenience.
- Customers can transact with the organization as they wish, at their desired time, and from a desired context (especially in the case of web-based transactions).
- Opportunities for service failures decrease as there is reduction in human intervention; like it or not, humans are a major cause of process failures.
- It eliminates wastes in process.
- It drives cost efficiency.
- It gives the customer a sense of control.
- There is no need to wait to speak to an executive from the customer's perspective.
- It helps to reduce the customer's interaction time with the organization.

However, organizations need to know that whatever self-service it provides to customers has to be engaging and value adding. If the customers do not see the self-service options as value adding, the entire endeavor is going to be a waste.

The following are some of the examples of self-service options that organizations have provided to customers:

- ATM for cash withdrawal from a bank
- Mobile phones for utility payments
- Kiosks for airline check-in
- Web check-in from home
- Interactive voice response (IVR) in a call center

Self-service strategy can be driven with an objective to gather either cost efficiency or customer convenience. My recommendation is that a Lean implementation should focus on both.

However, we need to remember the following while creating a mechanism for self-service for the customers:

- Do not make self-service an objective of Lean process improvement. It should emerge as one of the solutions.
- Before providing self-service solutions, complete a thorough value-added/non-value-added analysis.
- A customer self-service solution should not appear as something that has been pushed by the company but something that the customer needs.
- Find the customer segment that is being catered by the self-service solution.
- Focus on providing the right customer experience.
- The self-service experience has to be the same or better than other channels.
- Make sure that the customer is able to navigate easily through the channel.
- Buttress the self-service product with marketing and customer communication *blitzkrieg* if need be.
- Constantly listen to the customer: Be ready to continually change or improve the self-service channel based on customer requirements.
- Provide training to customers so that they become comfortable using the self-service channel (incentivize customer service agents to do it, such as in a call center when a customer does not know how to use an IVR system).

- Incentivize customers to adopt the self-service channel.
- Leverage superior technology wherever required to provide better customer convenience.

From my experience, self-service is a brilliant solution when you have a large number of customers to take care of, such as for a bank, airlines, travel agency, and so on. The self-service channel should solve some problems that the customer is facing today, or else one may find it difficult to create an impact. But remember, if the ease and convenience is affected, the customers can return to the old way of interacting with an organization and not adopt a self-service channel. Last but not the least, do not push the self-service channel. Give the customers a choice and engage them to adopt it for their personal betterment.

42

Encourage Team Members to Report Problems

We were facilitating a Lean breakthrough or a Kaizen Blitz in a process at a back office of a financial services company. The shop was run by a woman who was supposed to be very successful in the organization. She had a command-and-control leadership style and believed in a "no-questions-asked" style of management. The team members were expected to execute like robots. The processes were threaded with metrics, and the employee performance was measured by health of processes as indicated by the metrics. Their increments and bonuses were dependent on meeting the metrics as set up by the operations head. She took a lot of pride in having created a "disciplined" workplace where employees carried out all instructions to a T.

After the Lean breakthrough, it came out that there was much waste in the process and that the process efficiency was less than 1%. When the final presentation of the findings was made to this woman, she blew up. She rebuked the process owner on how he had so far been running a process that had been laced with wasteful activities. Instead of encouraging the team on the wonderful job in identifying inefficiencies that were tacitly hidden in the crevices of the process, she undid their efforts by verbal repartees.

This is an example of how a leader can undo the efforts of Lean. Do you think people who were involved in this Lean breakthrough would ever be comfortable with doing a Lean project? They would never, ever dare to come out with the wastes that may be impeding the process performance. They will hesitate to share the problems and deficiencies of the workplace, as nobody wants to be rejected for efforts that are not respected.

There is an important lesson to be learned here. Leaders implementing Lean should work toward creating a culture where their team members come and report their problems. They should encourage their teams to regularly come and share the process, workplace, and business concerns. As a matter of fact, leaders should be worried if teams do not come and report the problems on a regular basis. Yes, this is even if the processes are performing well and have no customer issues.

An organization implementing Lean should encourage problem reporting and abnormality identification. Leaders have to shun the command-and-control leadership style, which kills messengers of bad news. In such a scenario, the teams will never come up with new ideas. This is foundational to problem solving, which is so integral to Lean transformation. A command-and-control leadership style does not work in today's world. You may obtain some results in a short time, but I doubt if they will sustain over a long period. Command-and-control leadership stifles innovation and impedes all bright individuals who want to positively contribute to the company's success. Can we afford this today when product life cycles are becoming shorter and technology has been a force multiplier? With a large number of opportunities available today, an organization will find it difficult to retain these individuals. A command-and-control leadership style also fosters self-protection: Individuals try to protect themselves from blame, reprimands, and criticisms. This is a sheer waste of organizational time and resources that builds avoidable inefficiencies. If the leaders are serious about Lean implementation, they have to make these fundamental changes or else it shall result in a false start.

Remember that when you have an aware workforce, problems will be reported. This is the best thing that can happen to a company. It is always better to report the problems than let them hide in the organizational crevices only to create bigger problems in the future. Let the message go around the firm that problem reporting is a way of life. Train all your employees on problem-solving skills so that they not only report problems but also are involved in solving them. If they are not able to solve them, they seek help from others.

43

Processes Should Positively Affect the Key Stakeholder

We all know that processes are an integral part of product or service delivery. Processes comprise input, output, procedures, tasks, and policies. Creating standardized processes is necessary for all service businesses. The standard processes have to be installed across all service delivery centers to ensure consistent delivery of product or service to the customers.

However, when we talk about a process, we only keep customers in our mind. Our sole objective is to create a good process so that it meets the requirements of the customer it serves. There is also a lot of focus to ensure that the process meets the needs of the business. Organizations spend a lot of energy to achieve this objective, for which teams take up projects whose success is celebrated with a lot of hype.

However, we often seem to forget a key stakeholder the processes serve. Beyond the customer and the business, the processes need to serve the people who run them. In our urgency to meet the other requirements, we seem to forget what a process should do or be for the individuals who work on it. Remember, it is the people who give life to the process, so meeting their requirements is the key. While designing processes, it is imperative that they should positively affect all three key stakeholders: customer, business, and people.

A good process should have the following impact on the people who run it:

- Makes process execution easy
- Is easy to follow
- Has smooth flow and minimal or no back-and-forth action or loop-back
- Is easy to adopt and adapt for ongoing execution

- Helps to get on board and train the uninitiated
- Helps to get more from fewer resources
- Breaks down silos and builds cross-functional coordination
- Improves team efficiency and effectiveness (efficiency means something helps deliver more from less; effectiveness means something helps to meet the objectives of the team)

A process improvement that positively affects only one or two of the three stakeholders can be suboptimal, and the fullest outcome from the process may not be achieved. This is also a reason why designing and improvement of a process should always happen with the involvement of the teams who run it.

Table 43.1 is a short questionnaire that leaders can administer regarding their processes to see if they have been designed or improved to positively affect all three stakeholders. Answering yes to all the questions will mean that the process is suitably designed or optimized.

TABLE 43.1

Deb's Questionnaire to Determine Effectiveness of Process Optimization or Design

Serial Number	Pointers	Yes/No
Customer: The Recipient of the Product or Service Produced by the Process		
1	Is the process helping to meet the requirements of the customer (requirements comprise cost, speed, quality, safety, convenience, etc.)?	
2	Is the process making it easy for the customer to interact with the organization?	
Business: The Larger Entity That Houses the Process and Works toward Meeting Certain Objectives		
1	Is the process making a positive impact on the business objectives or workplace performance?	
2	Is the process the most efficient and cost-effective way of accomplishing the results?	
People: The Individuals Who Are Involved in Running the Process		
1	Is the process supporting the performance of the people who run it?	
2	Is the process making it easy for the team to run it, and was it designed and implemented with their involvement?	

44

Do Not Forget to Ascertain the Health of Lean Adoption

Just beginning with Lean implementation may not be sufficient. Companies start with a big bang, and after some time, the deployment loses steam. While derailment is always obvious, it makes sense for one to keep the focus on things that are not happening as desired. However, this requires deeper insights on the nuances of Lean implementation and the relevant change dimensions around it. This approach can be subjective and person driven. This is where a diagnostic tool such as that shown in Table 44.1 can be of help.

I would recommend that leaders spend 15 minutes each quarter to see if their journey is under way as desired. For large organizations, this tool can be used by the leaders in each of the locations or geographies. The tool is so handy that leaders across hierarchies can use it.

Once you have completed the tool, use Table 44.2 for interpreting the results.

In the early days of the journey, you will never find everything in place. However, with a well-thought-out road map, you will know exactly when the journey would be arrested.

The Lean diagnostic health instrument may not be comprehensive but does give a quick peek into where your deployment is heading. It will let you know the lever that needs to be pressed to obtain the desired success.

TABLE 44.1

Deb's Lean Health Assessment Tool

Serial Number	Pointers	Yes/No
1	There is another competing initiative that is being run in the company.	
2	You see major political resistance (internal or external) that the senior management cannot, or would not like to, overcome.	
3	You see Lean not being driven as a strategy to take the performance of the company to the next level but as another flavor of the month.	
4	You do not see members of the top management team taking time to drive the Lean transformation.	
5	The required resources have not been provided for the success of this intervention.	
6	The rollout of Lean is not getting the required visibility in the company.	
7	The outcomes of Lean are not threaded to the performance management of all key employees in the company.	
8	Approach to project selection is ad hoc and does not follow a business rationale.	
9	The initiative is largely being run by the team members of the Lean office without participation of the front-line team.	
10	The measures of success have not been defined clearly.	
11	Teams are being pushed into projects without proper skill development.	
12	Teams are branding cost-cutting initiatives as Lean.	
13	The Lean rollout is happening with the mandate from the top but with no inputs from the employees.	
14	Implementation of Lean is project based, and there is no focus on the anchors.	

TABLE 44.2

Deb's Lean Assessment Tool: Interpretation of the Scores

Number of Yes Answers	Interpretation
>7	Major cause of worry. You need to see if you have conceptualized your rollout correctly. It also puts a question mark on your Lean sensei or the person who is guiding the company on Lean efforts.
5–7	It is worrisome. Find out the reasons for the gaps.
Less than 5	May not be a cause of great worry, but do find out the reasons.

45

Embed a Regime of Reflection

As a part of Lean thinking, embed a culture of "reflection" in your company. After every project or assignment is completed, make your teams ponder and unfurl on what could have been done better. Install a process to find out what went well and what did not.

The Japanese call this process *Hansei*, which literally means self-reflection. It is an integral part of Japanese culture: An individual acknowledges one's mistake and commits to carry out the necessary improvement. The Japanese society constantly pushes its politicians to carry out Hansei and mend their corrupt ways.

In Toyota, teams conduct Hansei events on a regular basis, such as during Lean breakthroughs or Kaizen events. The outcomes of the Hansei event are looped back to improve products or processes. Taiichi Ohno used to say that Hansei is Check when referring to the PDCA (Plan-Do-Check-Act) cycle.

Beyond Japan, the US army has a brilliant process of reflection that they call "After Action Reviews" (AARs). They are among a few organizations in the world to have institutionalized this process. The AAR is an assessment discussion to improve the effectiveness of teams. The objective of the AAR is to identify what was supposed to happen, what actually happened, why it happened, and how to sustain strengths and improve on weaknesses.

As you begin Lean implementation, create a learning organization by practicing Hansei on a regular basis.

So, how does one go about installing a good reflection process? Figure 45.1 shows the process to follow.

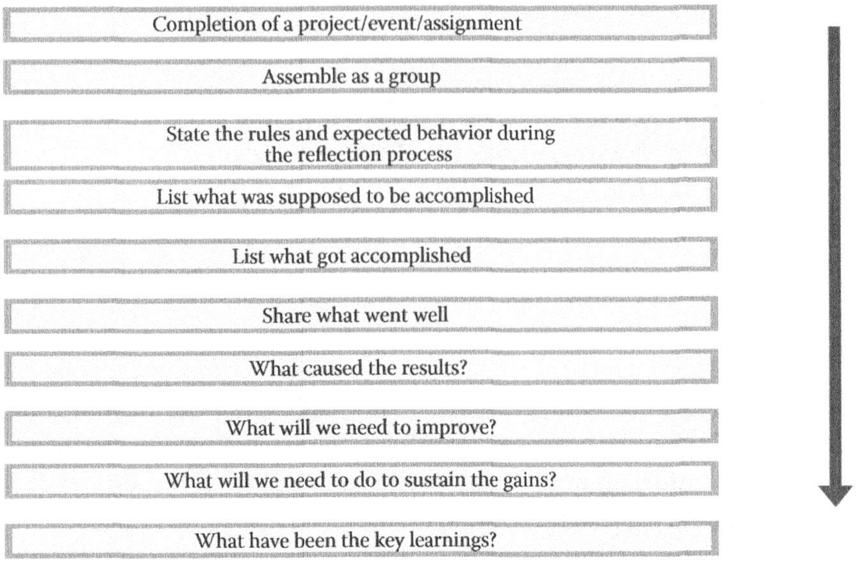

Completion of a project/event/assignment

Assemble as a group

State the rules and expected behavior during the reflection process

List what was supposed to be accomplished

List what got accomplished

Share what went well

What caused the results?

What will we need to improve?

What will we need to do to sustain the gains?

What have been the key learnings?

FIGURE 45.1
Process for reflection.

The following list summarizes the do's and don'ts of a reflection process that you always need to remember:

- Never criticize a person.
- Capture the war stories and lessons learned.
- Probe why certain actions were taken.
- Discuss in sufficient detail so that everyone understands.
- Make juniors lead the discussion.
- Do not lecture and sermonize.
- The reflection process can happen before, during, and after the event.
- Can be used in any process or context for which there is repetitive work.
- This should bring out the training needs of individual team members.
- Use open-ended questions.
- Involve all participants in the discussion.
- Focus on both individual and team performance.
- Have the reflection as soon as the project is completed.
- Make sure everyone is on board before completing the reflection event.

Reflection has to be both an intellectual and an emotional exercise. It helps the team members to dissect the entire project or assignment in a nonthreatening manner. When you start the reflection process in your company, it will take time for members to open up. Leaders have to be patient and should create an environment for the members to open up. All individuals in the company should undergo training on how to facilitate a process of reflection.

Even if the project has been successful, do carry out the reflection process. There will always be learnings, opportunities for improvement, and sustainability challenges to discuss. It is a brilliant engine to make tacit knowledge explicit.

46

As You Negotiate the Lean Journey, Do Not Forget Those Who Could Derail the Efforts

Embarking on a journey of Lean deployment is not a bed of roses. Just picking tools from the quiver and deploying them does not mean things are done. Despite thinking tools have been deployed in the right manner, many deployments fail because there are individuals who may not have come on board and have not put their hearts and minds in the Lean effort. A Lean professional should constantly keep his antennas on to find who is supporting the Lean efforts and who is not. One of the traits of successful Lean deployment is that people are engaged in the Lean journey and are committed to achieving the shared objective of the transformation. Smelling and ascertaining resistance are key roles of a Lean professional. This is a skill that a Lean professional should acquire and polish over a period of time. I have seen many Lean professionals who are adept regarding tools and techniques. They not only know them but also know how to apply them successfully. But, where they falter is in their ability to understand who are the resistors in the transformation and if employees are engaged in the journey.

This is not easy and requires patience and a "be-at-it" mindset to find the resistance on the horizon. Like any change endeavor, resistances could be of two types: active and passive. This is summarized in Table 46.1.

I would like to caution all Lean professionals keen on driving Lean. Do not declare early victory. Table 46.2 summarizes a few instances when commencement of Lean has not delivered benefit to the fullest. As a person driving Lean, one needs to be careful not to land in such a situation. But, if you go through the list, many of these instances will

TABLE 46.1

Active Resistance and Passive Resistance

Active Resistance	Passive Resistance
These are resistances about which teams are very vocal and readily share all their concerns. These resistances are easy to manage as one can exactly address the concerns of the impacted employees.	There are quite a few resistances about which teams do not share. They happen by inertia: The teams do not adopt what is being pushed. Because they do not share their concerns, it becomes difficult to address such resistance. Passive resistance can quietly derail the whole movement, and the Lean Change Agents need to be sensitive to this. Proactive efforts have to be made to show hidden concerns that teams may share concerning Lean implementation.

TABLE 46.2

Challenges and Resistance in Making Lean an Integral Part of the Organizational Fabric

Name	Details
The dissonance at the peak (Figure 46.1)	This refers to a state wherein the top management does not have its mind in Lean implementation. The chief improvement officer pushes an agenda of Lean, but the chief executive officer (CEO) and his direct reportees do not support it to the fullest. The CEO may give a patient hearing but does not spend time subsequently to get the agenda moving or getting his reportees to adopt this practice.
The dissonance midway (Figure 46.1)	This refers to a state wherein the engagement is confined to the top management and the teams at the grassroots level. The middle management is unknowingly kept out or just partially engaged in the implementation of Lean. Also, this occurs when middle management has limited incentive to adapt to this practice.
The dissonance at the foothill (Figure 46.1)	This refers to the state wherein there is no or partial engagement of the teams in the front line or process associates. There are tall claims and a lot of hullabaloo by the leaders, but the teams below them are not involved in the transformation. Large projects are done that involve the top and middle management, but teams below these levels are not involved. Many times, the leaders look at process associates as robots who have to keep their mouths shut and work.
The siloed drive	In this case, the implementation of Lean is confined to a few functions and departments. However, the overall organization does not reap the benefits as the processes improved are not end to end and the employee engagement does not have a critical mass.

(*Continued*)

TABLE 46.2 (CONTINUED)

Challenges and Resistance in Making Lean an Integral Part of the Organizational Fabric

Name	Details
The lost ground	This happens when implementation is done, but over a period of time the results wane. The leadership team becomes aligned, a large number of improvement projects are done, awareness sessions are held for the employees, but over a period the benefits recede. The key reason is a lack of infrastructure for sustaining the gains.
The quiet onslaught	This happens when there is an eerie silence among all the employees across hierarchies (especially middle management and front liners). The employees are indifferent to the Lean adoption as they believe that it is another of those "flavors of the month" that will come and go. There is a lot of skepticism in employees' minds because of past experiences of taking up similar initiatives that do not lead to a logical closure. This also happens when employees think that Lean will lead to loss of jobs.

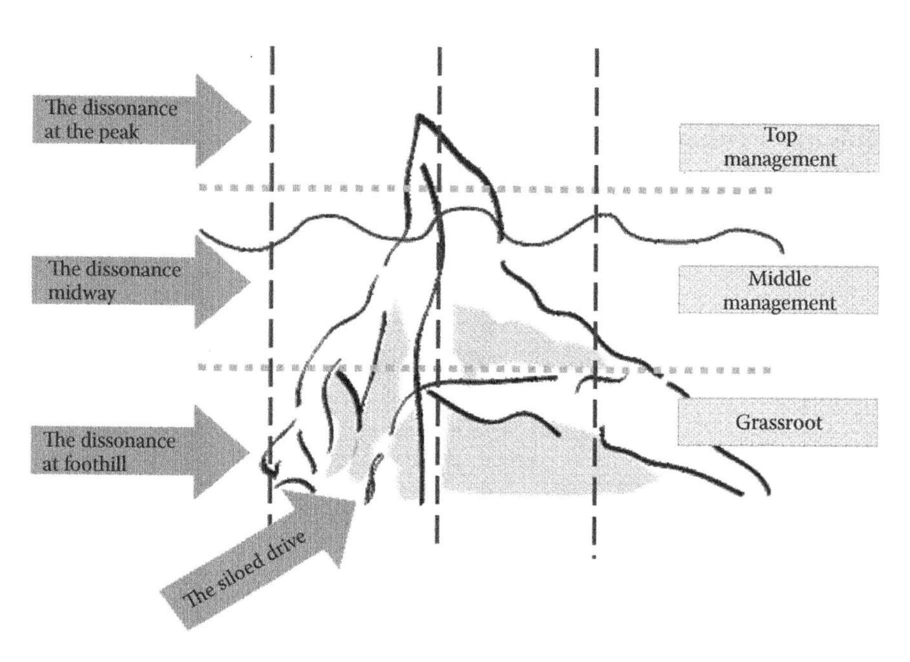

FIGURE 46.1
Visual depiction of resistance.

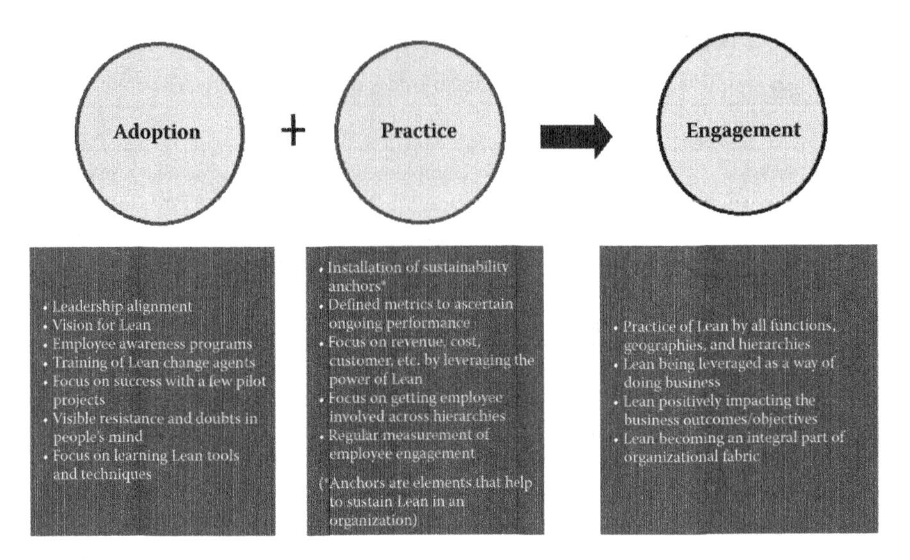

FIGURE 46.2
Adoption to engagement: the Lean approach.

be unpleasantly familiar. These are real-life cases I have experienced myself.

Just adopting Lean is not sufficient. It has to morph into an enterprise-wide engagement. This happens when adoption enmeshes with practice. This is best summarized in Figure 46.2.

47

Not Only Visual Tools but Also a Holistic Visual Management System

Making workplaces visual is an integral part of Lean deployment. However, let me tell you that workplaces will not be visually managed just by wishing it. Teams will not be able to adopt and practice visual tools for performance excellence merely by being aware of visual management. What leaders need to know is that we need a holistic approach to installing and institutionalizing a visual management system. For this, an approach has to be put together by the Lean professionals and kept in mind as the Lean journey progresses. Together with this, leaders have to create an environment in which teams are encouraged to come up with various ideas for visual tools that can be used in managing workplace performance. These tools will change as the journey of Lean matures and different interventions come to play. The worst thing is when someone reads about a set of tools in a book and simply implements them indiscriminately in the workplace. The type of techniques used should be driven by the type of Lean intervention under way. You will see that the visual management tools change, are replaced, and also improve over a period of time. The tools have to be relevant to the context. All the visual techniques in a workplace have to work in tandem to meet the larger organizational objectives. Figure 47.1 shows an approach that can be kept in mind while instilling visual management in an organization. You need to remember that this approach has a service industry bias, but a similar one could be devised for a manufacturing company also.

It may be noted that the visual tools shown in Figure 47.1 are not comprehensive. The objective here is to give a glimpse into some of the relevant tools at relevant points of the Lean transformation based on the initiative under way. Also, the various initiatives shown in Figure 47.1 may not be

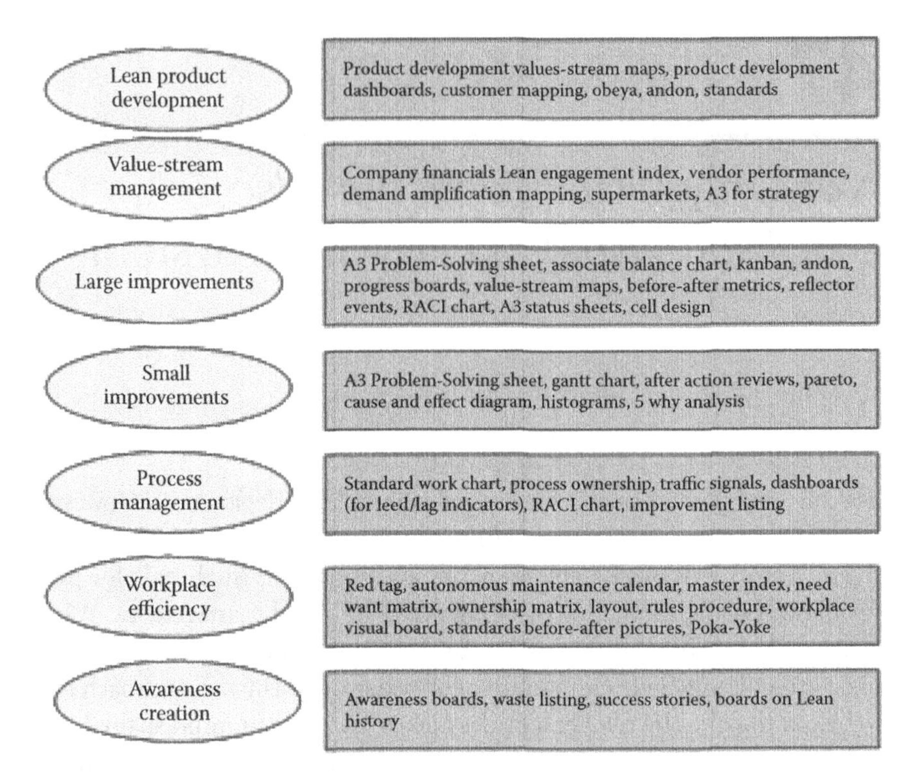

FIGURE 47.1
Approach for visual management.

sequential as depicted. Organizations can have these interventions in a different sequence.

As the organization migrates up the chart or matures in its Lean journey, some of the foundational visual tools may remain while others may be replaced with newer ones. For example, in the early days of Lean implementation, when awareness creation is the objective, we could inundate the workplace with posters on Lean basics and wastes. However, once a process management system is built, many of the posters on Lean basics and wastes could be replaced with standard worksheets, metrics, dashboards, and so on. It should be kept in mind that visual management tools have two broad purposes. They are used for information sharing and for managing performance. It is critical that a workplace should have a balance of both types. On average, 40%–45% should be information sharing, and the rest should be for management and improvement of performance.

Visual techniques need not necessarily be only visual but could be enmeshed with audio signals, like a siren. You also need to keep in mind that the types of visual management used during a project would be quite different from the ones that are used for day-to-day management of performance. The former would typically be like a war room, while the latter will be in the workplace where the day-to-day actions happen.

I would recommend appointing a person who can manage visual management in a workplace. This is a part-time role, but this individual keeps leaders informed on the health of the visual system. Of course, the person is only a provider of tools and techniques; the work has to be done by the teams. In fields such as financial services, use of information technology could be essential, especially when processes happen in electronic space and cut through geographies and functions. Remember that visual management displays can never hide in your respective computers. Even if they are digital, they have to be visible to all affected in a workplace.

48

It Helps to Ascertain Effectiveness of Visual Management

A key aspect of Lean management is visual management. Visual management is an approach to create an information-rich workplace, using visually stimulating aids such as graphical displays, boards, symbols, electronic boards, digital boards, lights, and so on, with an objective to manage and control performance. It is a proactive approach that translates critical information into visual forms that cannot be ignored. The important data of the company are converted to sensory stimulating information that not only can sustain the current performance levels but also can help to facilitate continual improvement. The larger objective of visual management is to drive operational excellence, cut wastes, and enhance processes.

A good visual management system should enable teams working on the shop floor to take real-time actions on processes. Within a short period of time, an individual walking around the process would be in a position to know what is going on in the process, the current process performance, and which actions should be taken to ensure the products or services are right the first time.

The tools used for visual management are targeted toward the following types of information:

- Strategic information
- Tactical information

Table 48.1 provides the difference between the two types of visual information.

As far as possible, visual management should be a low-cost solution. However, in service companies, technology is often leveraged to create solutions that may require investment, such as digital dashboards. Whatever approach is used, it should make the processes and workplace

TABLE 48.1

Difference between Strategic Visual Information and Tactical Visual Information

Strategic Visual Information	Tactical Visual Information
These are data that provide the strategic view of the company. They help the associates obtain a larger picture of the firm's performance. This information includes strategic parameters such as volume, profits, operating income, costs, and the like.	These are visual tools used to manage the day-to-day operations of a workplace. This information includes metrics on various operating parameters, such as process efficiency, people productivity, process lead times, and so on.

TABLE 48.2

Deb's Instrument for Assessing Visual Management Effectiveness

Serial Number	Attributes of a Visual Management System (VMS)	Score (1–5)
1	Has been designed and created by the employees of the workplace.	
2	Echoes the health of the workplace: Walking around the workplace clearly tells you what is happening there.	
3	Is intuitive, which means that employees are able to take action just by looking at it.	
4	Clearly tells the teams the standards that need to be met.	
5	Facilitates quick recovery when processes go awry (i.e., defects are produced or correct processes are not followed).	
6	Guides and tells teams what they need to do to accomplish their day-to-day jobs.	
7	Aids in identifying wastes or inefficiencies in the system.	
8	Customer demand is visible to the process associates.	
9	Product or service being processed gets pulled by the customer.	
10	Is a low-cost solution and did not require any investment.	
11	Helps to find abnormalities in the workplace.	
12	Provides "real-time" information (read: metrics) on the workplace performance (includes people, process, resource utilization, production, etc.).	
13	Does not allow defective items to enter the next stage of the process.	
14	Facilitates solution of day-to-day problems in the workplace.	
15	Used by leaders to review progress on a regular basis.	
16	Refreshed on a regular basis and teams use proactive questions and change it for its relevance.	
17	Has a live workplace visual board that acts as an information center for the workplace wherein teams meet at least twice a day.	

transparent. The whole objective of visual management should be to make the workplace or process speak for itself.

So, how does one go about deploying a visual management plan? Lesson 47 provides details of the approach. The only thing that needs to be remembered is that it should be treated in a holistic manner. This means that leaders need to have a comprehensive plan while creating an environment for the teams to develop it in an organic manner.

The question that often bothers us is, How do we know that a visual management system has been designed correctly? Teams put a lot of effort in installing visual management controls but are not sure if they have done all the right things. An instrument shown in Table 48.2 will help you ascertain the quality of your visual management system. Score each of the parameters and see the health of your visual management system. The rating scale of 1 to 5 has been used, and this is detailed in Table 48.3.

The interpretation of the scores can be performed using Table 48.4.

Remember, a good visual management system has to be made with the involvement of teams. It has to be so engaging that it catalyzes teams into workplace improvement.

TABLE 48.3

Visual Management: Rating

Rating	What Does It Mean?
5	This is best in class and a benchmark for all to emulate.
4	This is very good and has become institutionalized across all teams.
3	This is functioning, but there is only partial adoption.
2	This has been installed, and awareness has been established.
1	This has just been installed or begun.
0	This does not exist.

TABLE 48.4

Visual Management: Interpretation of Scores

Score	Interpretation
Above 75	This is a world-class visual management system that drives workplace efficiency and continual improvement.
55–74	The visual management system has begun to act like an engine for workplace improvement.
35–54	The visual management system has been partially deployed.
20–34	The visual management system has just been established.
Less than 20	There is no visual management system existing in the company.

49

Ohno's Wastes Are Applicable to Service Organizations

Many Lean practitioners tell me that the eight wastes of Lean as advocated by Taiichi Ohno are not applicable in service companies. This is not true. Kudos to the work done by Mr. Ohno: These wastes are applicable across businesses in all contexts and situations. Table 49.1 lists the eight wastes of Toyota with specific examples from service businesses.

The key here is to identify the unseen wastes. This is not easy, especially because individuals bound to a workplace tend to ignore what is happening there. I call it the "ostrich syndrome."

Ostrich syndrome is a phenomenon in which individuals are unaware of the loads of waste that they are sitting on. Most organizations, over a period of time, become a victim of the ostrich syndrome. The larger the organization, the bigger the likelihood of this problem.

The problem sets in further when the company is growing at a fast rate and leaders find it futile to spend time on unfurling wastes. The only way to reduce and arrest this problem is awareness among employees and coaching them to see wastes in their respective workplaces. This is a major change exercise in an organization's Lean transformation as it requires employees to do so voluntarily.

TABLE 49.1

Eight Wastes of Lean with Examples

Type of Waste (Symptom)	What Does It Mean?	Examples
Waste of Overproduction	Processing too much or sooner than required.	• Purchasing items before they are needed • Processing paperwork before the next person is ready for it • Franked loan agreements lying unutilized • More promotion materials printed than required
Waste of Motion	Unnecessary movement of individuals for successfully completing a job in a process.	• Multiple visits by salespeople to obtain the right documents from customers • Scattered departments in an organization • Zigzag movement of documents during the processing of an application form in a process shop • Walking to and from a copier, central filing, or fax machine while executing a process
Waste of Inventory	Refers to items or supplies in the process that are more than a single piece, flow, or movement of items (which is under process) one at a time in a process. In a service business, because a single piece may not be possible, it is about supplies or items that are more than required.	• Documents lying in a workplace because they have not been lifted by courier on time • Piles of loan files lying in branches or credit shops • More stationery than required • More information technology equipment than required in a workplace • Documents or records lying beyond their retention period
Waste of Transportation	Refers to movement of more materials than required. (Note: Waste of Motion is about "movement of people.")	• Excessive e-mail attachments • Multiple handoffs • Multiple approvals • Files moving from one branch to another

(Continued)

TABLE 49.1 (CONTINUED)

Eight Wastes of Lean with Examples

Type of Waste (Symptom)	What Does It Mean?	Examples
Waste of Waiting	Refers to individuals and items being idle between operations	• Customers waiting in a bank branch or grocery store • Files and documents waiting for signatures or approval • Associates in a process waiting for preceding processes to finish • New employees waiting for infrastructure or a computer • Customer waiting in a phone banking or call center queue
Waste of Underutilized People	Refers to employees in a process not being utilized to the fullest	• Employee talent not being leveraged • Individual put in the wrong job • Incorrect selection of a candidate for a job
Waste of Defects	Refers to waste that occurs due to errors and not getting the processing right the first time	• Error in filling a cell phone application form • Incorrect name placed on a checkbook
Waste of Overprocessing	Refers to efforts in a process that add no value to the customer	• Sending a document by overnight courier service when an ordinary courier service would suffice • Redundant steps in a process • Multiple inspections in a process • Lack of operator training • Undefined or unclear customer requirements

50

Are You Aware of Wastes of Business Acquisition?

When you work on Leaning your sales and marketing function, you should remember that there could be a hidden problem: burgeoning acquisition costs about which you may not be aware or know about very late. Traditional accounting systems do not treat customers as acquired assets, and costs associated with them are lost. Because accounting systems are primitive, a large number of cost overheads may not be allocated in the right manner.

As I have always believed, the cost of customer acquisition is a key metric for sales and marketing performance. Tracking this metric can be a good indication of the health of Lean application in sales and marketing.

Table 50.1 summarizes the 12 wastes associated with customer acquisition that I have found during Leaning of sales and marketing functions or processes. Although meant for service businesses, these wastes are fairly universal in nature and can be used in all types of businesses. One may ask about the eight wastes of Toyota discussed in Chapter 49. Well, these wastes are universal, but having specific wastes around customer acquisition helps the sales and marketing colleagues relate better. Or else, the debate can always be that Toyota's eight wastes are not applicable in sales and marketing functions. Having wastes concerning customer acquisition can be a great help. Learning not to see them will only add to customer acquisition costs.

Identifying the wastes in business acquisition is the first step toward making them Lean. This may appear to be a tall order as the sales and marketing functions are traditionally thought not to be amenable to process thinking. Also, the general belief is that friends from sales and marketing

TABLE 50.1

Deb's 12 Wastes of Customer Acquisition

Type of Wastes	What Is It?	Additional Details
1 Waste of product or brand inundation	This is a waste associated with launching more products than required.	This is often done with an objective to acquire additional customers. Often, this is without keeping the customer needs and requirements in mind. Mindless addition of products not only stresses the businesses but also breeds unwanted complexity.
2 Waste of misaligned distribution channel	This is the waste associated with not leveraging the right distribution channel for the targeted customer segment.	The distribution of products and service has to be done in a cost-effective manner by leveraging the right channels. The focus should be to improve acquisition efficiency, meet the specific needs of the customers, and achieve better penetration. For example, a bank may need to use low-cost channels such as the Internet to target the unbanked segments.
3 Waste of atrophied skills	This is a waste associated with employees not having the desired skills to serve the customers.	This happens when sales colleagues are not adequately trained and lack the fundamental skills to manage long-term relationships.
4 Waste of hazy customer cohorts	This is the waste associated with not unfurling the unique needs and requirements of customers within existing customer segments.	Within customer segments, there could be specialized customer groups organizations need to cater to. For example, in the financial services domain, a mass-affluent segment could have unique needs of customer groups such as retirees, investors, baby boomers.

(*Continued*)

TABLE 50.1 (CONTINUED)

Deb's 12 Wastes of Customer Acquisition

Type of Wastes	What Is It?	Additional Details
5 Waste of customer ownership	This is a waste associated with inadequate ownership of total customer relationship.	In many companies, the customer is handled by a large number of functions so no one really owns the customer. All that happens is that the ownership of the customer is passed on from one function to another. For example, in a cell phone service provider, the customer is acquired by the sales team, after which the details are passed on to the customer service teams for ongoing issues. These customer service employees just address the service issues, and there is nobody to manage the customer relationship.
6 Waste of inadequate existing customer focus	This is the waste associated with not leveraging the existing customer relationship for furthering business.	This happens when companies do not focus on acquiring the share of the wallets of existing customers. Or, even if there is a focus, the current process of cross-selling is not effective. For example, a bank should build a relationship to offer products at various life events of the customer.
7 Waste of coordination	This is the waste associated with follow-ups and back-and-forth actions that happen in the acquisition process.	These are typically associated with efforts spent on cross-functional coordination.
8 Waste of dissonant partnership	This is the waste associated with not leveraging the potential partners or entities beyond the company to further business (or sales).	This happens when organizations do not thread a relationship with partners to gather and serve the customers better and profitably.

(*Continued*)

TABLE 50.1 (CONTINUED)

Deb's 12 Wastes of Customer Acquisition

Type of Wastes	What Is It?	Additional Details
9 Waste of inflexibility	This is the waste associated with a firm's inability to embed flexibility into its products or services.	This waste can be avoided when flexibility is built into product design to allow creation of products at short timescales. This can be achieved when organizations modularize product components so that new products and features can be assembled based on changing demands of the customer.
10 Waste of wrong selling	This is a waste associated with selling the wrong product to the customer.	This happens when the focus of the organization is not on providing profitable solutions to the customer but somehow to sell whatever is available to them.
11 Waste of missed opportunities	This is the waste associated with being unable to fathom the emerging opportunities of the customer.	This happens when the company is proactively not able to sense and respond to the changing needs of customers, competitive pressures, and market dynamics. This requires creation of capability within the firm to meet precise customer demands amid changing market conditions. Also, organizations need to identify growth opportunities for newer products or services.
12 Waste of missing processes	This is a waste due to lack of processes within the sales and marketing function.	To address this problem, organizations need to treat business acquisition as a process and embed with a large number of metrics to track its performance on an ongoing basis.

are too busy to get business and are not bothered about the hidden inefficiencies. This is an incorrect view.

If you are able to show the sales colleagues the hidden wastes in their processes, they would be the first people to adopt them, as it helps not only to make their processes efficient but also to meet the customer demands and also in a cost-effective manner. The intervention of Lean should help to focus the company around profitable customers and building a profitable relationship.

So, where does one begin, and what are the few things that an organization should do to instill Lean thinking in sales and marketing processes?

Here is a list of dos that you may like to look at to build a Lean customer acquisition process:

- Train all sales and marketing colleagues on the 12 wastes of customer acquisition.
- Do the value-stream mapping of all sales and marketing processes to unfurl the hidden wastes.
- Create flexible platforms with modularized components to generate newer products at short notice.
- Build an engine within the company to sense and respond to changing customer demands.
- Leverage potential partners to serve the customer in a cost-effective manner.
- Scan existing market segments to find out the customer cohorts within them.
- Align channels with products to facilitate revenue maximization in a cost-efficient manner.
- Unfurl the unmet needs of the customer.
- Ensure that the business acquisition professionals have adequate weight to acquire new customers and increase the share of wallets from the existing customers.
- Focus on a mutually beneficial relationship.
- Focus on the lifetime value of customer relationship.
- Use analytics to target appropriate customers.
- Optimize existing processes.
- Track the following metrics:
    - Trend of overall acquisition costs
    - Channel wide acquisition costs
    - Return on investment of marketing costs

    - Market share
    - Rate of customer acquisition
    - Lifetime value of customers
- Clearly enunciate when to draw a line with customers—all relationships are not profitable.

Remember, Lean thinking is about not acquiring a large number of customers but focusing on profitable customers.

51

Be Careful about the Service Recovery Process

Many customer service professionals would be astonished with this chapter. Their view would be, How can we have a service organization and not talk about the service recovery process? So much has been written on the service recovery process and how it can help to get back customers. Complaint handling and service recovery go hand in hand and are looked at as an enabler for customer retention. Research has shown that customers whose concerns are resolved quickly are more likely to be loyal to the company. Efficient service recovery helps to elevate the organization's capacity to maintain an effective customer relationship. Service recovery also provides an opportunity to the organization to further its business by cross-selling or up-selling its products and services.

It is not that I am against service recovery processes. They are required and must exist in all businesses. If customers are discomforted, the onus is on the company to make the required corrections so that customers can be retained.

My discomfort is not with the service recovery processes, but with the goal of an organization to build good service recovery processes and protocol. As Lean practitioners, we should realize that service recovery is about addressing the symptoms and not getting into the root cause. We really do not realize that just focusing on service recovery is associated with one of the following:

1. It adds to unnecessary costs. We do not realize that installing service recovery protocols means putting in place an infrastructure to manage them.
2. If we just focus on service recovery, we could be forgetting to prevent the defects that could be happening in the process.

So, what should be the approach to this whole issue in an organization on a Lean journey? The following are a few pointers that should always be kept in mind:

- Make defect prevention the goal of a process.
- Do everything in the process and train the associated people for better customer engagement.
- Design processes that ensure the output (product or service) is flawless.
- Focus on providing such impeccable service that customers do not come back to you with their problems.
- Provide world-class training and capability development programs.
- Provide proactive service instead of reactive service.
- Make everyone in the company own service, not just the customer services department.
- Create an engaging service that customers enjoy.

Once these tasks have been done, the company should put in place a world-class and robust service recovery process that helps to manage the service failures. This process should be so good that even if the customer is peeved, the customer will decide not only to stay with the company but also to advocate others to do the same.

However, it would be suicidal for an organization to focus only on building service recovery process and not focus on designing processes and building people who can create flawless products and service. To see if companies are getting this concept right, companies should keep their eye on the metrics shown in the following list. Over a period of time, the percentage of service recoveries should decrease.

- Number of products or service delivered right the first time
- Percentage of products or service delivered right the first time
- Number of service recoveries
- Percentage of service recoveries

52

Multiskilling Is a Good Capacity Optimization Technique

A key solution implemented during a Lean intervention in an organization is multiskilling of employees. Multiskilling or cross-training is done to achieve the following objectives:

- Create capacities
- Manage peaks in demand or volumes
- Distribute capacity
- Create competency
- Bring the need for capability building of process associates to the agenda of top management
- Optimize the process
- Build process flexibility and resilience, which helps to manage emergencies and exigencies, such as a 9/11 event

So, how does one go about multiskilling in a service context? The approach that follows is one that I have found useful.

STEP 1: PROCESS SELECTION

Select a process, a process family, or a set of processes based on a certain criterion. It could be based on volume, business objective, or cost.

STEP 2: OBJECTIVE IDENTIFICATION

Define the objective of the process or business targeted for multiskilling. Look at all the business objectives that it needs to accomplish. Specifically, look at the productivity numbers that are being targeted.

STEP 3: VOLUME DISSECTION

Do a thorough demand analysis. Look at volumes and the trend of customer requirements. Obtain data on the following:

- Average demand
- Peak demand

STEP 4: EXISTING NUMBERS

Take stock of the existing manpower numbers and sections where you would like to improve productivity.

STEP 5: PROCESS DECONSTRUCTION

Break the processes into activities and tasks (Table 52.1).

TABLE 52.1

Processes Broken into Activities and Tasks

Process	Activity	Tasks

TABLE 52.2

Example of Activity Prioritization for Multiskilling

Process	Activity	Importance
Account opening process	Mailing room	1
	Inwarding	3
	Verification	10
	Risk assessment	7
	Scanning	10
	Processing	3

Note: Use a scale of 1, 3, 7, 10 for importance rating. Importance can be based on business objectives, productivity, or customers.

STEP 6: ACTIVITY PRIORITIZATION

Among all the activities, further prioritize the activities that you may like to take in step 1. Ideally, all the activities in the selected process or process family should be examined for multiskilling. This is done as sometimes it may not be possible to take up all activities for multiskilling due to resource constraints. As an illustration, please examine Table 52.2, which provides an example of activity prioritization for multiskilling.

Please note that in each of the activities, there could be a large number of tasks for individuals to work on.

STEP 7: SKILL COMPETENCY MAPPING

Map the existing skill levels of all employees and grade their competencies as shown in Table 52.3.

TABLE 52.3

Gradation of Skills

Grade	Details
Novice	Uninitiated and yet to be exposed to the activity or task
Understudy	Currently undergoing training or has completed the training and currently is implementing the same; needs continual supervision
Practicing	Can independently do the activity and does not require supervision
Expert	Has mastered the activity and can train others

STEP 8: IMPLEMENTATION PLAN

Put in place an implementation plan. Ensure that the progress is reviewed by a senior leader on a regular basis. Given the organic nature, it may make sense to roll out the cross-training in phases. The broad approach to be followed is discussed next.

Phase 1

Focus on +1 and −1. This is about cross-training individuals on skills preceding and succeeding an activity of a process. This should happen in a period of 3 months.

Phase 2

Focus on cross-training all associates within value stream, business, or operational silos. This could happen in a period of 6–12 months.

Phase 3

Focus on cross-training relevant associates in at least three core processes or process loops in a selected value stream. This should take a time frame of 12–18 months.

Phase 4

In this phase, look at cross-training all relevant associates in all core processes or process loops. This should take a time frame of 18–24 months.

TABLE 52.4

Dashboard for Tracking Multiskilling

Metrics
Number and percentage of associates multiskilled in a process
Number and percentage of members equipped on +1 to −1
Number and percentage of core processes that have multiskilled associates
Time taken to get the new process up with the cross-trained resources
Time taken to allow a process to become operational with the multiskilled resources

STEP 9: INSTALL METRICS

To ascertain the ongoing performance, a few metrics should be put in place that should be tracked through a dashboard (see Table 52.4).

Remember, cross-training takes time and can happen when it is driven strategically. Do not expect quick results. You have to be patient and keep at it. It is only then that you will see it as a lever for business enhancement.

53

Building a Pull System in a Service Enterprise

The concept of pull is essentially about responding to a customer's demand. The objective of a Lean system should be to design such products or services that meet the changing needs of customers. The customer could be the end customer (consumer or end-user) or even the next person in the process. The objective of a pull system should be to provide the customer what he requires, when he requires it, at the price he is willing to pay.

Simply put, the principle in *pull* is about not producing anything before the next step in the process. In a larger scheme of things, pull is about orchestrating with all stakeholders in the process (such as vendors, outsourced partners, etc.) with an objective to provide the customer a solution to his problems at his convenience and at the right price.

Examples of a pull system are as follows:

- Dell will not build a computer until you order one.
- Checkbooks are made after customers request them through the Internet.
- Kanban cards trigger an order of stationery items in an office.
- Supermarkets are not storing more than what is required and replenishing items on a daily basis.
- Music companies are not pushing their products (songs on CDs) to retail shelves, but customers are pulling their desired content from Amazon or iTunes.
- Rather than pushing individuals into classrooms, companies have created a large number of e-learning content that can be accessed by employees when needed (clearly a pull-based approach).

However, creating a pull system that connects the entire end-to-end process can take time. One has to proceed step by step. This becomes especially difficult when processes are large and cut across geographies. In such a case, it is recommended to break the processes into subprocesses and then connect them through a pull system. The first step should be to streamline and create continuous flow in the subprocesses. This would entail removing bottlenecks from these processes and ensuring that the items processed move without any hurdle. Having smoothed the subprocesses, the focus should be on connecting these subprocesses through a pull system (see Figure 53.1).

In an ideal case, you would like the entire end-to-end process to have a continuous flow. But, in real life you will, in all likelihood, have to create a hybrid model that has both pull and push concepts.

A push system assumes customers are mute spectators who will accept whatever you offer them. In a pull system, a platform is created so that resources can enter to provide customers what they need, when they need it, and in the quantity they need. In a pull model, distributed resources

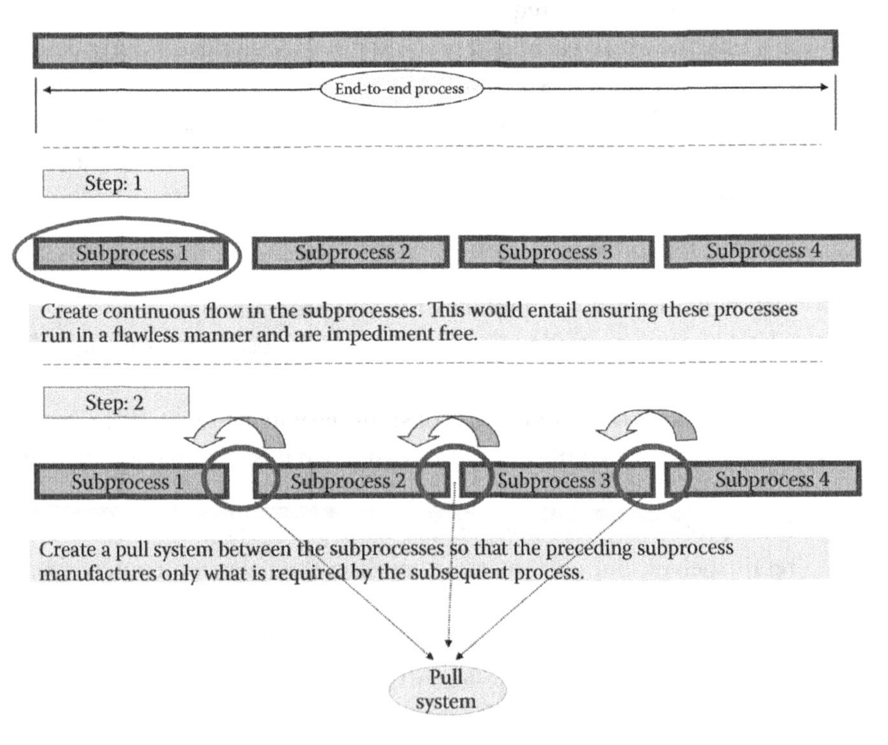

FIGURE 53.1
Stepwise approach to creating a pull system.

come together to work on the unanticipated needs of the customer. The platform in a pull system should be able to provide the diverse needs of customers. The platform here refers to a mechanism that allows a broad range of capabilities to come together. A push system comprises tightly coupled modules that function in a predefined sequence. A pull system contains loosely coupled modules that are actually designed to build flexibility.

These modules could be subprocesses, physical entities, vendors, and so on. Based on the customer requirements, which could often be unanticipated, these modules come together to create the product or service. An important dimension of a pull system in a service context is distributed communities coming together to find innovative ways to solve customers' problems. One of the objectives of Lean transformation is to make sure that there is an impeccable relationship among the various parties in the network.

54

Know the Little-Known Law

There is a little-known law that has many applications in Lean solutions. It is called Little's law. Derived by John Little at the Massachusetts Institute of Technology Sloan School of Management in 1961, this law has universal application to all types of business systems. All leaders and practitioners involved in Lean deployment should know this law clearly and understand its application.

Little's law states that the average number of works in progress in a stable system is equal to the completion rate multiplied by their average time in the system. This relationship is as follows:

$$\textbf{Work in progress} = \textbf{Throughput time} \times \textbf{Average completion rate}$$

This relationship can also be shown as follows:

$$\textbf{Throughput time} = \frac{\textbf{Work in progress}}{\textbf{Completion rate}} = \frac{\textbf{Work in progress}}{\textbf{Input rate}} = \frac{\textbf{Work in progress}}{\textbf{Output rate}}$$

To understand this better, let us look at a back-office process shop (Figure 54.1) where the mortgage application forms enter and exit at a constant rate of 100 files/day. If the average inventory of application forms at any given moment is 800 forms, we can use Little's law to find out the throughput time or the average time that the files are within the shop. Using the formula, the throughput time will be 8 days.

Let us see some real-life applications, which will demonstrate the power of Little's law and how leaders can use it. Please refer to Examples 54.1 through 54.3.

So, as we have seen, Little's law has universal application. It is directional in nature, and using it successfully can enhance the efficiency of a company.

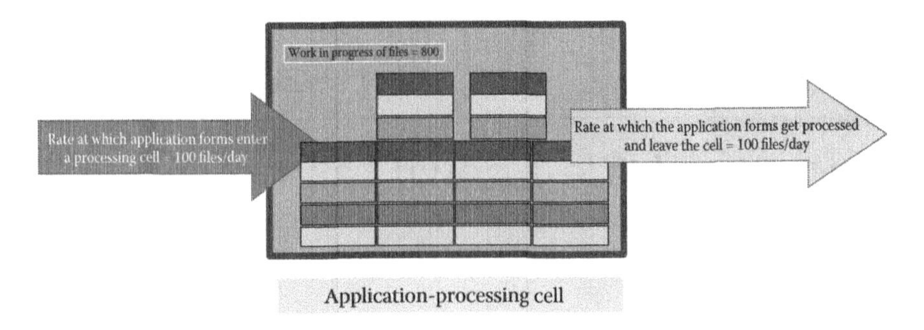

FIGURE 54.1
Back-office process shop showing entry and exit of application forms.

Example 54.1: Branch Teller

You are running a retail bank branch. Let us talk about the tellers used by customers for transactions. On average, 6 customers are serviced every 20 minutes, and there are 18 customers in the queue. The branch manager wants to know the average time a customer will spend in the line.

We could use Little's law to respond to the branch manager.

Here, the work in progress (WIP) is 12 customers.

Rate at which customers receive services = 6/20 = 0.3 customers

Average time customers will be in line = 12/0.3 = 40 minutes

Yes, it takes 40 minutes to service the customers. The bank needs to do something about this before customers switch to other companies.

Example 54.2: Mortgage Back-Office Processing

A mortgage back-office processing area receives an average of 320 applications per day. It takes around 30 minutes to process each application form. Each shift has 9 hours with a 1-hour break included. How many people are needed to process the application form?

In this problem, we need to find the WIP, which is people.

It is important to convert all to the same unit.

Throughput time is 30 minutes = 0.5 hour

Rate at which applications are received per hour = 320/8 hours = 40/hour

WIP = Throughput Time × Completion Rate

WIP = 40 × 0.5 = 20 people

Given the current throughput times, the processing requires 20 people.

It may be noted that as a part of Lean implementation, we may decide to improve the productivity of employees, which results in more applications processed by the same people, or we may decide to redeploy the existing employees.

Example 54.3: Inventory in a Store

A mom-and-pop store sells 700 kg of chicken every week. The owner wants to ensure that the stock of chicken is never more than 2 days old. What is the quantity of chicken that the owner should store to ensure that none of the inventory crosses the 2-day threshold?

Rate at which chicken is bought at the shop = 700/7 = 100 kg/day

Ensure all items have the same unit.

Throughput Time = 2 days

Average inventory that should be in the shop (WIP) = $100 \times 2 = 200$ kg

55

Use Little's Law to Create Pull in Transaction Processing

Little's law can come in handy for creating pull within processing cells. To create pull, the objective of a cell should be to maintain a stipulated work in progress (WIP; also called a calculated threshold) that has been reached based on the desired standard lead times and the processing (completion) rate. The endeavor should be to maintain the calculated WIP and restrict entry of items until the WIP in the cell is reduced below the calculated threshold. To explain this point, let us look at an auto finance credit-processing cell.

The following are some of the process data:

Average rate at which the application forms are processed per day = 100 (i.e., on an average 100 files are processed each day by this cell)

Standard throughput time with which the application forms need to be processed = 12 hours (0.5 day)

Let us use Little's law to find out the desired WIP to ensure the throughput time of 12 hours (0.5 day) is maintained. We incorporate the data into the formula of Little's law, which is as follows:

0.5 days = Desired WIP/100 application forms/day

Required WIP in the cell = 50 application forms

This means to meet the throughput time of 0.5 days at any given moment of time, the average WIP in the cell should be 50 application forms. The

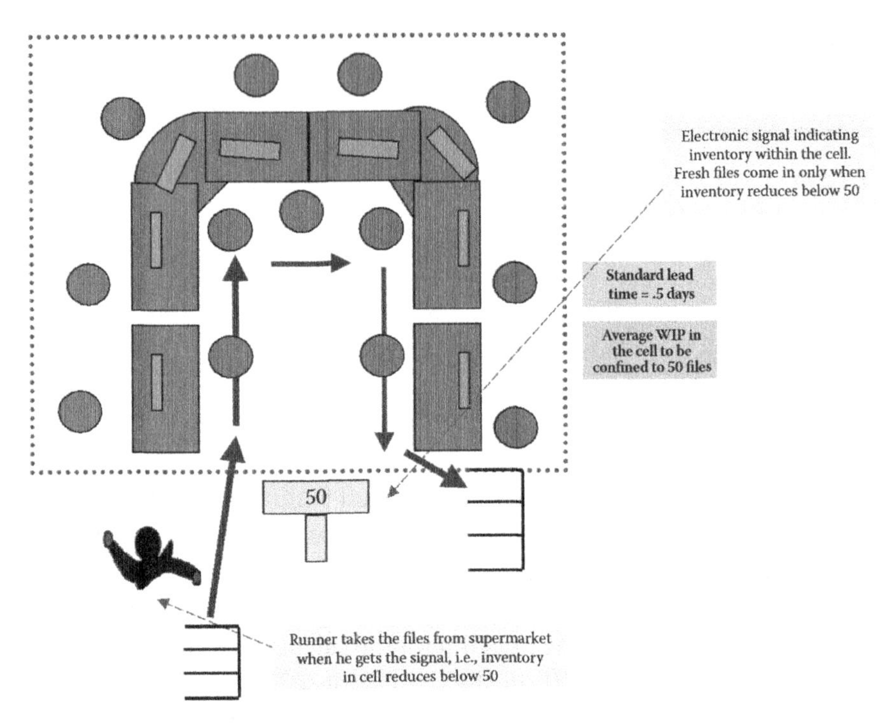

FIGURE 55.1
Stipulated inventory managed through an electronic signal in a credit-processing cell.

team should put into place a mechanism to ensure additional application forms do not enter the cell until the number of WIPs in the cell comes down to 49.

This can be achieved by an electronic signal as shown in Figure 55.1. The cell pulls in inventory only when the inventory falls to 49 when the runner replenishes the same from a supermarket.

There are practitioners who say that pull in a Lean context is not possible. They are not correct. There are a number of ways by which pull can be created in a service context. The example given is just one of them.

56

Do Not Standardize All Processes in a Service Company

Companies embarking on Lean adoption go ahead and initiate enterprise-wide rollout of the standard process. Practitioners of Lean often get standard work (Lean practitioner jargon for standardized procedure) done for all work that happens in the firm. You will see such procedures being displayed across desks of associates doing jobs such as data entry, customer service, billing, hospital surgical preparations, account relationship management, design, and so on. This is done more out of a senior-level dictate and not as something that would enable their work. Have we not seen those colorful standard works on desks? They look colorful but are rarely used.

Why does this happen? This happens because the person who is providing Lean leadership has not fully understood where the principles of standardization are relevant. Because Lean talks about standard work, all processes are documented in detail, with a belief that this will help in the Lean journey. This approach is myopic. We should never forget that the outcome of Lean has to make a positive impact on customers and not just standardization. So, it is imperative to understand where the principles of standardization can have a positive impact on the customer experience. The other day, I attended a conference wherein the global quality leader of a leading service company said that the company had standardized all their processes irrespective of the functions and context. When I asked him the impact of this on customer experience, he dodged my question and did not respond. The response did not surprise me.

Does it mean that in a service context, processes should not be standardized? Is it true that standardization generates waste in service processes?

Standardization is required but cannot be carried out in a mindless manner. Standardization in service businesses should be done based on the context and the complexion/type of processes that have been taken up for improvement. For example, in a retail branch, the "DD" (demand-draft) or "cash" process can be standardized as this process is repetitive. However, you cannot standardize a process that handles "queries." Because this process handles queries, a majority could be different. So, wherever you have processes that handle exceptions instead of standard queries, the teams should come up with a broad set of guidelines that customer services executives can use. Standardization here does not mean that the customer services executives will become robots. Each of these processes (whether standardized or not) will have to be backed up by highly competent and empowered customer services executives who are able to manage the interaction experience of customers. Of course, standardization will result in waste when you try it on a process that does not require standardization.

There are two approaches that can help you make a decision whether standardization is the right strategy during a Lean performance improvement journey.

APPROACH 1

Remember that there are two things that drive standardization: visibility and variability in customer requirements. Visible processes are those in which the customer is involved and the process acts on the customer. In a "nonvisible process," the information and material provided by the customer are worked on. In the former, quick response times and customer interaction skills are critical, while in the latter, there is a lag between customer request and delivery. The higher the visibility of a process, the greater the degree of variability in customer requirement and hence the greater the requirement for customization. Processes that require a higher degree of customization should not be standardized, while the ones that happen at the back end can be standardized. Figure 56.1 summarizes this concept. I would recommend that before embarking on a Lean transformation, all processes should be placed on a scale like the one shown in Figure 56.1, and then you can make the decision to standardize or not

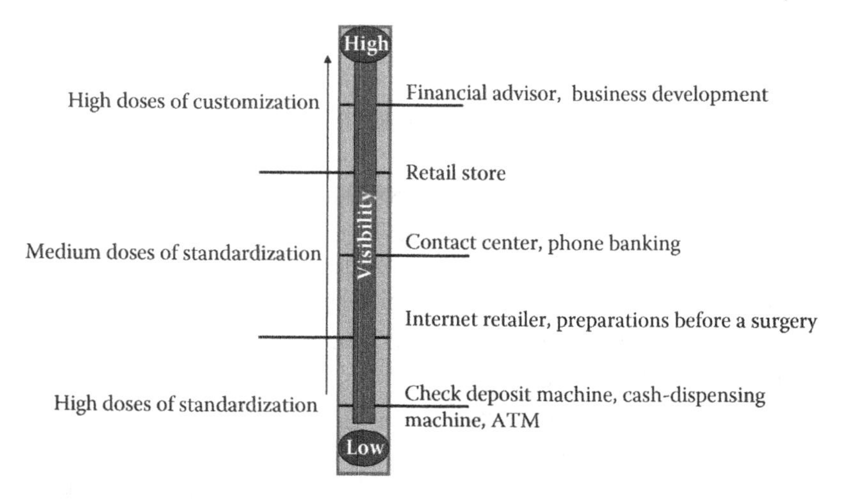

FIGURE 56.1
Process visibility and degree of standardization.

standardize these processes. So, what does one do when a process cannot be standardized? We should be able to provide guidelines for making judgments and decisions.

APPROACH 2

This is the better approach to decide whether to standardize a process. I would recommend that we use Hall and Johnson's approach. Joseph M. Hall and M. Eric Johnson came up with a matrix that tells you which processes should be standardized and which should not be.

On the y-axis of this matrix is the degree of variability in the product or service output of a process to a customer. On the x-axis, there is the degree of variability in inputs to the process. Looking at the variability of the output and input of a process, there are four types of processes before us. Let us look at each of them in detail as researched by Hall and Johnson:

Mass processes: These are processes that require low variability in process input and output. Here, the customer is looking for standard products or services for a narrow range of products or services. The focus should be to create standard processes that facilitate achieving these objectives. Examples of where such processes are required are automobiles, steel, aluminum, consumer durables, consumer financial products, and so on.

Mass customization: There are processes for which there is low variability of product input, and the customer values variability in the process output. Here, standard processes are required to produce a controlled variation in a product or service output. The best example in this category is Dell's personal computers, which cater to unique requirements of customers.

Nascent or broken processes: These are processes for which input variability to the process is high, and the process cannot create consistent output that the customer demands. These processes use new materials, design, or technology. Typically, these processes, like newborns, have yet to mature. The approach to deal with such processes could be either of the following:

- Because the output variability is high, see if customers can appreciate it. If they do, treat them as an "artistic process."
- If customers do not tolerate variation, the focus should be to arrest the variation and to create a standard process.

Artistic processes: These are processes that have high variability in input as well as output. Here, the customer values the variability in the products or services that are provided. Also, the inputs that trigger the process vary, making it challenging for the company. In such a case, focusing on reducing variation by standardizing the process is a wrong approach. The endeavor should be to provide a set of guidelines that allow exercising judgment and improvising the processes based on the context.

In a hotel, the guests come up with a wide array of requirements. In such a case, providing employees with standard work is not the right thing to do. Providing the employees a standard script, such as one that says, "Look into customer's eyes. Say 'Good morning,' 'Good afternoon,' 'Good Afternoon,' using words such as 'Certainly,' 'It's my pleasure,'" may not be fully effective. This is because each customer contact is unique and would require unique responses. Instead, providing employees with directional guidelines can be more helpful. This could include statements such as the following: "I will build an engaging relationship with customers," "I will look at each customer contact to delight the customers," or "I will ensure that customers' stay at the hotel becomes memorable." An approach like this ensures that the unique needs of customers are taken care of and the responses are based on what the customer wants and the customer's moods or emotions.

Other examples of artistic processes for which standardization should not be used but only guidelines should be provided are the following:

Customer contact center
Writing code for new software development
Relationship management with wealthy customers
Business development
Auditing
Customer service
Leadership training

In the context of application of Lean to a service business, understand the complexion of the process in the process matrix and accordingly make a decision.

To summarize, do not mindlessly standardize all service processes. Look at the context and make decisions.

57

Make the Process of Cross-Selling More Efficient and Effective

Cross-selling is a big focus of organizations. Having closely watched a large number of cross-selling processes in service organizations, I have seen how companies inadvertently generate waste, which can easily be avoided. *Cross-selling* refers to the approach by which you try to sell more products or services to your customer.

In an enthusiasm to increase business volume, organizations often implement a large number of cross-selling goals around their products or services. The belief here is that an increase in the company's share to a customer's wallet will lead to a long and profitable relationship with the customer. While this assumption may not be wrong, the existing approach for cross-selling could lead to suboptimal results.

Companies share their cross-selling figures with analysts and celebrate successes concerning them. It is possible that we may be celebrating something that may not really be achievements. While the volumes may be going up, demonstrating penetration and an increase in market share, the customer requirements may not really have been met. This is where Lean principles come into play.

Figure 57.1 shows the typical approach to cross-selling followed by companies. As you can see, the approach is inwardly focused. The focus is on peddling what is available and not really on what the customer values.

Redesign the process and keep in mind the voice of the customer and what the customer values. Create a process wherein the products are pulled by the customer. The new pull-based process is shown in Figure 57.2. The conceptual approach is explained in Figure 57.3.

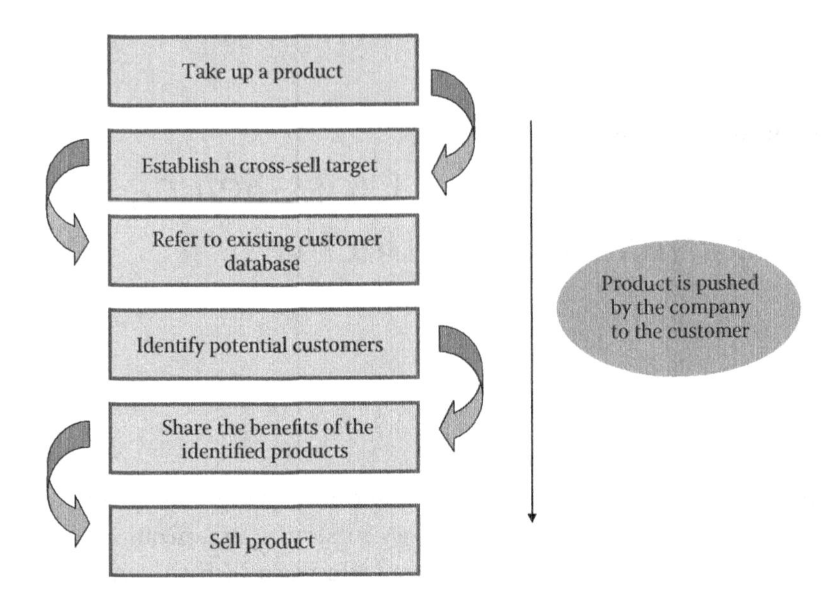

FIGURE 57.1
Typical push-based cross-selling process.

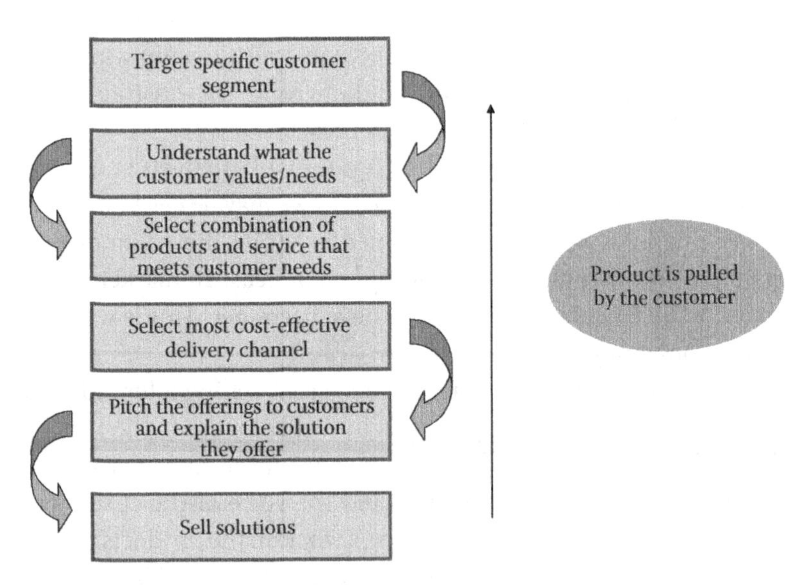

FIGURE 57.2
New pull-based (Lean) cross-selling process.

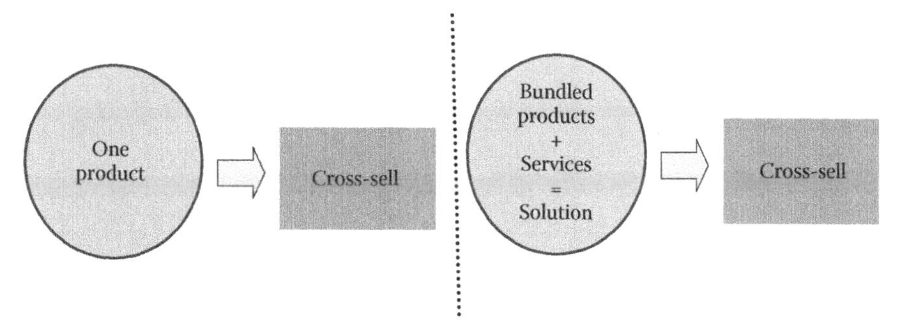

FIGURE 57.3
Philosophical difference between normal cross-selling versus Lean cross-selling.

The key traits of the new Lean cross-selling process are as follows:

- The organization provides a solution to a customer's need (remember, Lean is about solving customers' problems).
- The solution is pulled by the customer instead of being pushed by the company.
- The focus is on not only the product but also the associated services.
- Sales colleagues associated with the product are not only salespeople but also individuals who have been trained on multiple products and solution selling.
- There is a huge focus on assessing what the customer needs.
- Focus on mutual value adds: The customer gets what he wants, and it is economically rewarding for the company.

The objective of a Lean intervention should be to enhance cross-selling's effectiveness in an efficient manner. To put it simply, this is about expanding sales volume at a minimal cost with the existing customers. However, there is a need to ensure that the needs of the customers are met. There has to be a positive co-relationship between cross-selling and actual fulfillment of customer needs.

A Lean cross-selling process will increase not only the share of the wallet but also the market share. The former is about making the same customer spend more on products or services from the company, while the latter is about getting more customers. When a company follows the Lean cross-selling process, the overall cost of ownership decreases.

58

Practice Pull-Based Sales

Many service organizations employ sales partners to whom they outsource the sales processes. The purpose for which such entities are appointed is to drive sales to meet targets. While there is nothing wrong with gunning to achieve sales targets, the mindset and approach behind it are incorrect.

The whole engagement is driven by a push mindset by which the intent is to somehow sell (read: push) the product or service to the customer. As a result, products or services that customers do not want are dumped on them. This leads to wrong selling and customer dissatisfaction and creates an impression of an overbearing organization.

In an organization pursuing Lean, this approach has to change. Push-based selling has to metamorphose into pull-based selling. This will happen when the company creates an environment where customers come back, demanding more. This in turn happens when the role of the sales partner changes from being a peddler of products and services to that of being a relationship creator. Names such as direct sales agency, direct marketing agency, or distribution partners have to undergo a change and should be redesignated as customer relationship agency, relationship management partners, or strategic partners. This is not easy and would require the entire leadership to look at these entities differently and the way their performances are managed. The profile of individuals who work for these entities also has to be different. These entities need relationship associates, not just sales associates.

The typical impact of push-based selling by partners leads to the following inefficiencies:

- Low conversion ratio
- High customer attrition rate
- Problems of incomplete application, leading them to go back and forth

- Focus on completing the transaction at any cost
- Instances of wrong selling of products and services
- Little effort to understand what the customer really needs
- Customer defection after some time or after experiencing the first product or service

Pull-based selling requires the role of these business-gathering entities to change as summarized in Figure 58.1.

The purpose with which these partners have to be set up is to build a relationship that creates pull among customers to buy the company's product or service. The focus should be to achieve relationship primacy by demonstrating that they care for customers and provide customers with products or service relevant to them. Remember, this will happen when customer relationship management is treated as a strategic priority.

The endeavor should be to target customer loyalty and get the greater share of the customer's wallet. This would require the company to invest in training for the partners so that they are equipped to align customer requirements and needs with the right products. The focus of the partners should be not only the transaction but also the creation of a comfortable relationship with the customer so that the customer provides with repeat business. It would be critical here that the right metrics are installed to track the performance of the partners. The metrics should track not only volume but also quality, service, and loyalty. Also, penalize partners if customers defect within 90 days of acquisition. Yes, the partners would have to have technology enabled. This is required for business intelligence and customer analytics with an objective of finding the detailed profitability

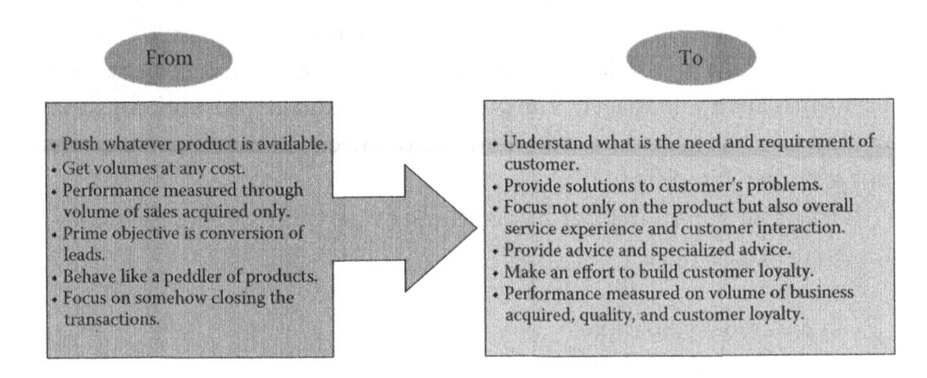

FIGURE 58.1
Role of sales partners in a company instilling Lean thinking.

and behavioral analysis of the customers to recommend better-fitting products or services. Of course, all these would require selecting the right partners, who have been closely scrutinized before appointment. After getting them on board, not only equip their teams with right skills but also embed robust processes.

A senior manager in the sales function should manage all these partners so that they are effective and efficient for the company. We may not realize it, but gaps in the partner relationship can create many tacit inefficiencies that can clearly be avoided.

59

Do You Know the DEB-LOREX™ Model for Lean Transformation?

Implementation of Lean in a service context is about intelligent customization of tools that have been applied to manufacturing organizations. Chief executive officers of service companies often ask me if there is a holistic model that will provide them direction on how to go about implementing Lean. My response is that the Toyota Production System is the model to look for. However, if you find it difficult to comprehend it as you may not have worked in service organizations, I would recommend that you look at the DEB-LOREX™ (Figure 59.1) model for holistic Lean deployment. It is my creation, and the model focuses on elements that are required to make a Lean service organization.

The components that make up the DEB-LOREX Management System (Figure 59.1) are as follows:

- Leadership
- Functions
- Value Streams
- Anchors
- Lean Thinking
- Results

> Are you wondering what the word *DEB-LOREX* stands for?
>
> It is an acronym for Deb's Lean Organizational Excellence Model. The word *Deb* is from the author's first name, Debashis. The words and abbreviations *DEB-LOREX model, DEB-LOREX Management System, D-LMS*, Lean Management System (*LMS*) are used synonymously—all mean the same thing.

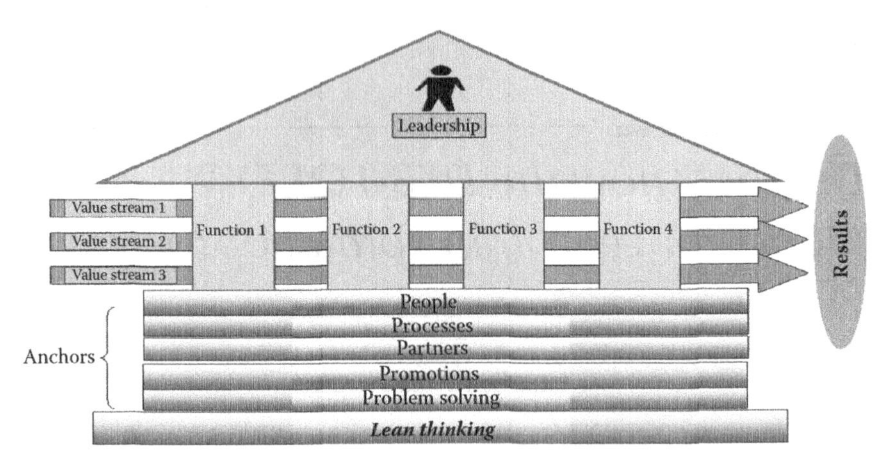

FIGURE 59.1
DEB-LOREX model of Lean transformation.

For Lean to deliver sustained benefits, it is imperative that all the components of the Lean Management System function in harmony. During Lean transformation, each of the elements of the DEB-LOREX management system must be properly implemented. Inadequacies in any of them will impair the overall expected performance of the organization. The elements comprising Leadership, Functions, Value Streams, Lean Thinking, and Anchors are the enablers in the DEB-LOREX Lean Management System (Figure 59.2).

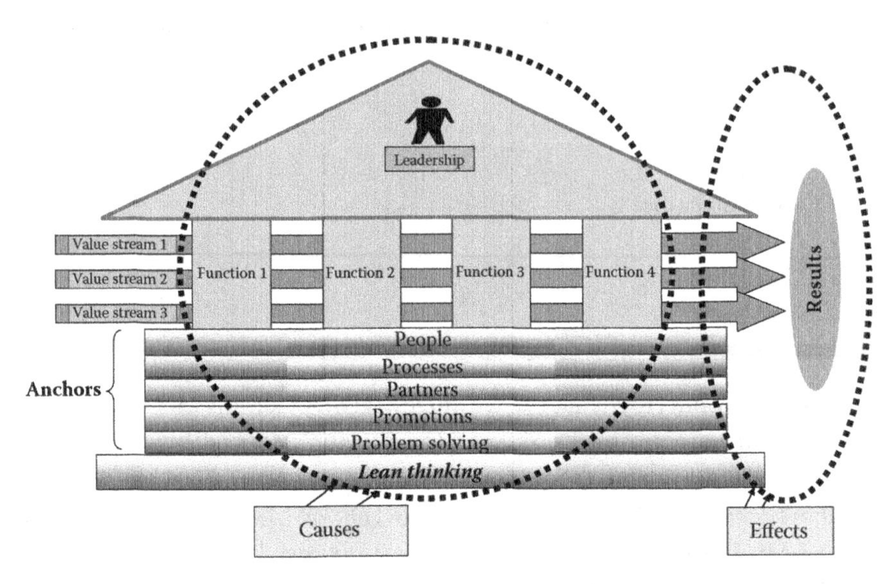

FIGURE 59.2
Cause-and-effect relationship of the DEB-LOREX model of Lean transformation.

The adequacy of each of the enablers will have a direct impact on the performance of the company. For example, an organization may have great processes, but without leadership commitment, it cannot expect the management system to deliver value. As the implementation of Lean matures in the company, leaders will be in a position to clearly say which of the enablers (or causes) needs to be acted on to obtain the desired results.

60

Make a Deep Assessment of Lean Enterprise Using the DEB-LOREX™ Index

In a previous chapter, I talked about an instrument to assess the health of Lean deployment. However, it is recommended that once a year you do a deep assessment to ascertain the holistic health of a Lean enterprise. It is a deep exercise and requires the involvement of key leaders of the organization. So, let me walk you through the details of the DEB-LOREX™ Index.

OBJECTIVE

The objective of this instrument is to provide an overview of the status and health of the Lean Management System (LMS) in an organization. It is based on the DEB-LOREX Management System, discussed previously in the book.

HOW IT IS TO BE ADMINISTERED

The instrument is to be used for a selected organization or business unit. The assessor should run through the points and grade them based on their current status. The assessment should be carried out by a qualified LMS assessor who spends sufficient time in the target business unit to carry out the assessment. The process of assessment comprises interviews, performance measurements, and observations.

FREQUENCY

While no specific time frames can be specified, a reasonable timeline has to be given to close the gaps observed from the last assessment. It has been seen that there should be a gap of a minimum three months between assessments, although one should administer it once a year.

WHAT DOES THE SCALE MEAN?

Rate each of the points on a scale from 1 to 5 where

5 = Strongly agree
4 = Agree
3 = Neither agree nor disagree
2 = Disagree
1 = Strongly disagree

Summary Scores

<table>
<tr><th>Serial Number</th><th colspan="2">Areas</th><th>Maximum Score (A)</th><th>Actual Score (B)</th><th>Percentage Score C = (B)/(A) × 100</th></tr>
<tr><td>1</td><td colspan="2">General</td><td></td><td></td><td></td></tr>
<tr><td>2</td><td colspan="2">Leadership</td><td></td><td></td><td></td></tr>
<tr><td>3</td><td colspan="2">Value stream</td><td></td><td></td><td></td></tr>
<tr><td rowspan="5">4</td><td rowspan="5">Anchors</td><td>People</td><td></td><td></td><td></td></tr>
<tr><td>Process</td><td></td><td></td><td></td></tr>
<tr><td>Partner</td><td></td><td></td><td></td></tr>
<tr><td>Problem solving</td><td></td><td></td><td></td></tr>
<tr><td>Promotion</td><td></td><td></td><td></td></tr>
<tr><td>5</td><td colspan="2">Customer</td><td></td><td></td><td></td></tr>
<tr><td>6</td><td colspan="2">Results</td><td></td><td></td><td></td></tr>
</table>

Overall LMS Score

Serial Number	Section		Weight in Percentage[a] (*W*)	Percentage Received (*C*)	Actual Score $D = (W \times C)/100$
1	General				
2	Leadership				
3	Value stream				
4	Anchors	People			
		Process			
		Partner			
		Problem solving			
		Promotion			
5	Customer				
6	Results				
Total			100%		
DEB-LOREX Index = Σ ($D1 + D2 + D3 \ldots Dn$) =					

[a] The weight of percentage of each section can be decided by the organization.

1. General

Serial Number	Points	1–5	Remarks
1	Lean to the company is about a holistic improvement journey that comprises making all the elements of the LMS work in tandem to achieve the larger vision of the company.		
2	The company is sensitive to the fact that Lean is about not only tools and techniques but also a philosophy for building operational excellence.		
3	Operational excellence to the company is about bringing about customer convenience, revenue enhancement, and cost efficiency and building a culture of continual improvement.		
4	There is a general belief among the employees in the organization that even the best of processes, workplaces, and business systems contain opportunities for improvement.		
5	Systems thinking is an integral part of the organizational fabric.		
6	The interdependencies and interactions among processes and the components of the LMS are known to all.		

(*Continued*)

Serial Number	Points	1–5	Remarks
7	The organization manages the components of the LMS in such a manner that it helps to deliver the vision crafted by the top leadership team and the board.		
8	All employees are aware that change in any of the components of the system will affect the performance of the overall LMS.		
9	There is a continual endeavor to improve the overall effectiveness of the LMS.		
10	The components of the LMS are managed in such a manner that people do the right things without being told.		
11	The LMS is not treated as separate from doing business; it *is* the business system.		
12	The operating teams clearly understand the cause-and-effect relationship between the components of the LMS and business results.		
13	The company uses this assessment checklist to ascertain the health of the LMS on an ongoing basis.		
14	The company has institutionalized the LMS assessment process into a positive, engaging process in which leaders at all levels become involved.		

2. Leadership

Serial Number	Points	1–5	Remarks
1	The Lean transformation in the organization is driven by the chief executive officer (CEO).		
2	The CEO and his or her direct reports are convinced that the LMS has the necessary wherewithal to make a positive impact on the performance of the company.		
3	A council comprising the CEO and his or her direct reports oversees the implementation of LMS.		
4	The leadership team of the company is using Lean as a strategy for business improvement and not just as another quality methodology to be used by quality project teams.		
5	The LMS council has set specific aspirational and future-state goals that need to be accomplished through the implementation of the LMS.		

(*Continued*)

Serial Number	Points	1–5	Remarks
6	The LMS council reviews the progress of implementation at least once a month.		
7	The organization has a vision, mission, and values that echo the principles of Lean thinking.		
8	The entire leadership team, comprising the CEO and his or her direct reports, understands the underlying principles of Lean and its key drivers.		
9	The leadership team demonstrates its commitment to the Lean transformation by voluntarily investing time whenever required.		
10	The leadership practices the principle of "customer first," which is about meeting the requirements of not only the end consumer but also the next person in the process (also called customers).		
11	The leaders know that successful implementation of Lean is about successfully adopting and practicing LMS across the company from top to bottom.		
12	The chief executive spends at least one day a month doing Ground Zero Walks to obtain a feel for the area where the action is.		
13	Leaders are driving Lean to bring in overall organizational excellence and not just to cut costs.		
14	Business leaders expect Lean to deliver a large array of benefits, such as revenue enhancement, service excellence, operational risk reduction, process efficiency, workplace safety, employee productivity, complexity reduction, and so on, over a period of time.		
15	The Lean transformation is being looked at as a change management intervention and not just another methodology for improvements.		
16	Each member of the leadership team and LMS council has participated in a Lean breakthrough.		
17	Leaders consistently seek to understand changing customer needs.		
18	The leadership team regularly participates in communication sessions not only to share the company's performance but also to energize the team to contribute to the Lean movement.		

(*Continued*)

Serial Number	Points	1–5	Remarks
19	The leadership team regularly communicates the importance of customer requirements and the role of the employees in making it happen.		
20	The CEO and the entire leadership team review the management report by the chief improvement officer and LMS office, which summarizes the overall health and status of implementation of the LMS.		
21	One of the areas that the company's leadership team emphasizes is building organization capability to sustain the Lean movement built around the LMS.		
22	All the leaders know that companies are a collection of people who voluntarily come together for a purpose, so they have to be engaged and not mandated into the LMS process.		
23	Leaders spend a lot of time coaching, mentoring, leading by example, and helping individuals to achieve their goals.		
24	The leadership team constantly focuses on creating a new generation of leaders who understand and drive the principle of the LMS.		
25	Leaders clearly know that piecemeal implementation of LMS will only deliver partial results.		
26	Leaders preach and practice the A3 framework for strategy deployment.		
27	Leaders refer to employees as associates and not as heads, bodies, or masses.		
28	Leaders at all levels know and manage collaborative groups for the successful implementation of the LMS.		

3. Value Stream

Serial Number	Points	1–5	Remarks
1	The organization is structured around value streams.		
2	Each of the value streams has well-defined ownership.		
3	The value streams encompass business units with profit and loss responsibility.		
4	The value streams have all the required capabilities to successfully carry out business.		

(*Continued*)

Serial Number	Points	1–5	Remarks
5	Employees at all levels in the value stream have performance management linked to outcomes of the LMS.		
6	Employees show a high level of engagement with the LMS.		
7	Value streams are shaped to serve specific market segments.		
8	The company has institutionalized a mechanism to determine the total costs of each of the value streams.		
9	The focus of the organization is value stream excellence and not functional excellence.		
10	The LMS office regularly makes an assessment of the overall waste in the value stream and shares it with all concerned.		
11	Each value stream has a Lean Maven working for them.		
12	Each value stream has its own leadership council to ascertain the progress of Lean implementation.		

4. Anchors

4a. People

Serial Number	Points	1–5	Remarks
1	Employees clearly know why the company has embarked on a journey of LMS deployment.		
2	Vision and Strategic Objectives are known to all employees.		
3	The entire leadership team, middle management, and bulk of the employees believe that people are the most important asset in the company and that they have to be treated with respect.		
4	The company places great emphasis on learning and development, and each employee spends at least 10 days on training that improves their effectiveness in their work and ability to work toward the larger vision of the company.		
5	The company hires employees who are sensitive to customer needs and shares the corporate values of the organization.		
6	Each employee in the company is evaluated by his or her superiors, peers, customers, and partners.		
7	Employees treat every customer interaction as an opportunity to make an impact.		

(Continued)

Serial Number	Points	1–5	Remarks
8	Messages conveyed by the workers are given due consideration by management for carrying out ongoing change in the LMS.		
9	Capability building of employees is looked at as a strategic initiative in the organization.		
10	All employees have been trained on problem identification and elementary problem-solving tools.		
11	Multiskilling of employees is taken seriously and reviewed by process owners on a regular basis.		
12	The company has a well-defined capability needs analysis process that is reviewed at senior levels.		
13	Employees understand that survival in the marketplace requires each one of them to contribute to making products or services right the first time.		
14	Employees are supported, not reprimanded, when they identify problems.		
15	Processes and procedures are designed with the participation of employees.		
16	There is a great amount of trust between the leaders and employees working on the process, shop floor, or workplace.		
17	Employees in a process regularly participate in improvements.		
18	Employees look at audits as opportunities to trigger improvement.		
19	The organization has a culture of problem prevention.		
20	Process associates take the lead to correct problems discovered online.		
21	Associates are empowered to take actions that facilitate quick customer recovery.		
22	The recruitment process endeavors to ascertain the current behaviors of prospective employees and how they will match up with organizational requirements.		
23	Each employee knows his or her customer and the end consumer and exactly what both of them expect.		
24	Employees proactively identify the barriers to meeting customer requirements and work toward eliminating these barriers.		

(*Continued*)

Serial Number	Points	1–5	Remarks
25	When something goes wrong in a process, employees discover the root cause of the problem.		
26	Employees practice value stream thinking, which is about sacrificing their personal and departmental concerns for value stream effectiveness.		
27	Employees have a high level of adaptability and quickly metamorphose with changes in customer requirements, technology, and competitive forces.		
28	Employees proactively look for wastes in their workplace or business and take the initiative to eliminate them.		
29	Employees actively collaborate with members of other functions and departments to solve business problems.		
30	When processes change, the employees quickly adapt to new roles and responsibilities with great agility.		
31	Regular feedback is solicited to ascertain employee engagement in LMS.		

4b. Processes

Serial Number	Points	1–5	Remarks
1	For each value stream, there are well-defined end-to-end processes.		
2	There is complete alignment on what comprises the value-creating processes.		
3	Clear categorization of processes into value-creating, value-enabling, and support processes.		
4	Detailed listing and inventory of all processes.		
5	All processes have been clearly defined, without leaving them open to interpretation.		
6	All processes have a defined purpose and objectives.		
7	Processes have defined standards to ascertain ongoing performance.		
8	Each process is backed up with procedures that help in their execution.		
9	For all processes, the potential risks have been identified.		
10	Users in a process clearly know the controls on the potential risks in the process.		

(*Continued*)

Serial Number	Points	1–5	Remarks
11	All end-to-end business processes have clear owners with the authority to design, maintain, and change the processes.		
12	All processes and procedures are adhered to as they have been designed and defined.		
13	Processes are managed using well-defined management processes.		
14	Process associates defuse identity of their own function, highlighting the identity of the process to which they belong.		
15	All key processes have metrics such as quality, delivery, cost, and customer service and business outcomes.		
16	There are instances when the organization fails to satisfy the needs of its internal functions but meets the needs of the customers.		
17	The sequence and interactions of the processes have been established.		
18	The functioning of the organization, business unit, or value stream is not affected if individuals such as value stream owners or key process stakeholders leave the organization.		
19	All processes are linked to policies that govern their functioning.		
20	Before processes are changed, the process owners always ascertain the likely impact on other processes.		
21	All process associates know the larger objective of the process and their role in making it happen.		
22	All processes and procedures are the best-known method of doing work.		
23	Processes have a number of visual indicators to make deviations obvious.		
24	Performance standards of the processes are known by the process associates working on them.		
25	All key business outcomes and Strategic Objectives are clearly linked and managed by processes.		
26	Process standardization is looked at as a first step to eliminate wastes from processes.		
27	Performance dashboards are displayed so that they are visible to all.		

(*Continued*)

Serial Number	Points	1–5	Remarks
28	All data collection for processes is automated and digitized.		
29	There is a hierarchy of dashboards so that people at all levels from CEO to process associate can see the relevant metrics.		
30	Takt time serves as a reference for designing all processes.		
31	All cycle times in the processes are standardized.		
32	An associate/operator balancing chart is used regularly to see how cycle times compare with takt times.		
33	Regular audits of the processes and procedures are carried out to ascertain adherence and reveal wastes.		
34	Standard processes are looked at as foundational to continual improvement.		
35	Process exceptions are virtually nonexistent.		
36	Intervention of technology happens in processes only after they have been Leaned and wastes have been removed.		

4c. Partners

Serial Number	Attributes	1–5	Remarks
1	Partners are treated as an extended arm of the organization.		
2	Trust is what drives the relationship between the company and its partners.		
3	The value stream owner becomes involved in choosing the partner.		
4	The company and the partner are in full alignment on organizational objectives and customer needs.		
5	The company believes that partners play a critical role in the success of the organization.		
6	The partnership strategy is clearly aligned with overall business and value-stream strategy.		
7	There are clear service-level agreements between the organization and its partners.		
8	Regular feedback is given to the partners on their performance.		
9	Regular training programs are conducted for the partner's employees to facilitate improvement in the partner organization.		

(*Continued*)

Serial Number	Attributes	1–5	Remarks
10	The company does not have a record of unceremoniously dumping partners.		
11	The decision on which partner to select is not based on cost but on an assorted set of value adds that it brings to the company.		
12	The organization regularly initiates collaborative projects and joint endeavors to get at root causes of problems.		
13	The company works with the partners to reduce the partnership cost.		
14	There is a constant endeavor to leverage the strengths and capabilities of both the company and the partners to meet overall organizational needs.		
15	The values and attitudes of the partner are important selection criteria.		
16	Both parties share their business strategies openly.		

4d. Problem Solving

Serial Number	Attributes	1–5	Remarks
1	Problems are looked at as an opportunity in the organization.		
2	Problem solving is looked at by all employees as a journey toward obtaining the best for the company.		
3	Each and every member of the organization is exposed to problem-solving tools and techniques.		
4	Problems are taken up for solution by all levels of the organization, comprising top management, middle management, junior management, and shop floor associates.		
5	The organization has the agility to quickly resolve problems and get at the root causes.		
6	The company has a management process to select the right problems to be taken up for resolution.		
7	Employees have developed a knack for problem identification, which they have been taught with relevant training.		
8	Leaders at all levels are concerned when problems are not identified in a process or workplace.		

(*Continued*)

Serial Number	Attributes	1–5	Remarks
9	Employees are encouraged and rewarded for identifying problems.		
10	The company has an approach for solving problems with the right quality methodology based on the complexity and type of problem statement.		
11	The company has a well-defined standard approach to ascertain the effectiveness of solutions implemented.		
12	Employees do not jump to solutions but spend adequate time understanding and defining the problem, followed by a structured approach to resolution.		
13	From the CEO to the janitor, every employee is familiar with the why-why analysis.		
14	For all problems taken up for resolution, the root cause analysis gets at the most fundamental reasons.		
15	Each problem has a well-defined action plan comprising what, who, when, and how.		
16	The effectiveness of solutions implemented is ascertained regularly by the LMS office and the value-stream owner and process owners.		

4e. Promotions

Serial Number	Attributes	1–5	Remarks
1	The company has a well-defined communication strategy for institutionalizing Lean across the organization.		
2	A marketing manager leads the communication and marketing of LMS philosophy.		
3	The organization's brand embodies all the elements of operational excellence that it is striving to achieve through the LMS.		
4	The LMS marketing team is successfully persuading the employees in the organization to adopt the LMS.		
5	Multiple channels of communication are being used to promote Lean within the company; these include meetings, intranets, brown bag sessions, road shows, events, brochures, merchandise, and so on.		
6	Everyone in the company is a brand ambassador for the LMS.		
7	There is an ongoing measurement to ascertain the effectiveness of communications.		

(*Continued*)

Serial Number	Attributes	1–5	Remarks
8	Rewards and recognitions are targeted toward all levels of the organization.		
9	The reward and recognition program emphasizes team performance while recognizing individual accomplishments.		
10	The rewards and recognition are designed to drive behaviors that are required for successful LMS implementation.		
11	The organization primarily focuses on nonmonetary rewards.		
12	The A3 framework and template are used by the entire company for problem solving and continual improvement.		

5. Customers

Serial Number	Pointers	1–5	Remarks
1	The CEO and leadership team believe that the organization needs to differentiate itself on customer service.		
2	Employees in the company know who their customers are in the process and also the end consumer that they serve.		
3	Retaining existing customers is a key focus area for the leadership team.		
4	There is a chief customer officer who represents the interest of the customer in the organization.		
5	Customer metrics are an integral part of the overall performance dashboard of the organization.		
6	There is an awareness in all process and value-stream owners of the impact of changes to processes on the customer.		
7	There is a well-defined management process to handle all the queries, feedback, and complaints of the customer.		
8	The company has a well-defined voice-of-the-customer strategy to ascertain the changing needs and expectations of the customer.		
9	The customer touch points in all value-creating processes have been identified, and customer listening posts have been installed in all of them.		
10	The organization has a service innovation cell to create service differentiation in organization.		

(*Continued*)

Serial Number	Pointers	1–5	Remarks
11	New processes are designed with the voice of the customer in mind.		
12	The organization has institutionalized an empowerment process that employees are supposed to follow when there is a service failure.		
13	The concept of customer retention is known to the bulk of the employees of the organization, and they demonstrate it in all their actions.		
14	The organization identifies specific areas in the customer experience that delight the customers.		
15	The leaders pay regular visits to customers to find out their concerns, problems, and headaches.		
16	The back-office team members meet with customers regularly to know their concerns, problems, and headaches.		
17	The organization has a process to weed out customers who are not profitable to the company.		
18	Customers are segmented to facilitate providing unique products and services.		
19	The company takes customer defection seriously and installs task forces to find out the causes for defection.		
20	The company solicits regular feedback from the employees on its products and services.		
21	The company works at creating employee and partner loyalty, as it believes that total customer loyalty is only possible when employees and partners feel loyalty to the company first.		
22	Achieving service reliability is a key objective of the leaders of the company.		

6. Results

Serial Number	Pointers	1–5	Remarks
1	The organization has a comprehensive dashboard for sharing the performance of financial numbers, customers, employee engagement, processes, partners, and people capability.		
2	The dashboard is digitized and captures data on performance at all levels of leadership.		

(Continued)

Serial Number	Pointers	1–5	Remarks
3	There is positive trending of financial results over the last 12 successive quarters.		
4	There is positive trending of customer results over the last 12 successive quarters.		
5	There is positive trending of employee engagement results over the last 12 successive quarters.		
6	There is positive trending of partners' results over the last 12 successive quarters.		
7	There is positive trending of people capability results over the last 12 successive quarters.		
8	The organization is meeting the performance targets for financial numbers over the last 12 quarters.		
9	The organization is exceeding the performance targets for financial numbers over the last 12 quarters.		
10	The organization is exceeding the performance targets for customer metrics over the last 12 quarters.		
11	The organization is exceeding the performance targets for employee engagement over the last 12 quarters.		
12	The organization is exceeding the performance targets for partners over the last 12 quarters.		
13	The organization is exceeding the performance targets for people capability over the last 12 quarters.		

Bibliography

Armour, Stephanie, "Who Wants to Be a Middle Manager," *USAToday*, August 13, 2007. http://www.usatoday.com/money/workplace/2007-08-12-no-manage_N.htm.

Bacharach, Samuel B., *Get Them on Your Side*, Platinum Press, Avon, MA, 2006, p. 6.

Chase, R. B., and Stewart, D. M., "Make Your Service Fail-Safe." *Sloan Management Review*, Spring 1994, Vol. 35, no. 3, 35.

Fujimoto, Takahiro, *The Evolution of a Manufacturing System at Toyota*, Oxford University Press, New York, 1990.

Gerstner, Louis V., *Who Says Elephants Can't Dance? Leading a Great Enterprise through Dramatic Change*, Collins Business, New York, November 2002.

Graban, Mark, *Lean Hospitals—Improving Quality, Patient Safety and Employee Satisfaction*, CRC Press, Boca Raton, FL, 2009.

Guarraia, Peter, and Andrew Schwedel, "For Banks in Need—Getting More from Lean Six Sigma," paper published by Bain and Company, New York, August 26, 2008. http://www.bain.com/bainweb/PDFs/cms/Public/FINAL-Lean%20Six%20Sigma Financial%20Services_ALL.pdf.

Hall, Joseph M., and M. Eric Johnson, "When Should a Process Be Art, Not Science?" *Harvard Business Review*, March 2009.

Hill, Linda A., *Being the Boss—The Three Imperatives for Becoming a Great Leader*, Harvard Business Review Press, Boston, 2011.

Hreniniak, Lawrence G., *Making Strategy Work*, Wharton School/Pearson Power, New Delhi, India, 2005.

Konz, Stephen, *Facility Design*, Wiley, New York, 1985.

Liker, Jeffrey K., *Toyota Way*, McGraw-Hill, New York, 2004.

Liker, Jeffrey K., and Michael Hoseus, *Toyota Culture—The Heart and Soul of the Toyota Way*, McGraw-Hill, New York, 2008.

Morgan, Rebecca A. "Getting the Laundry Out," Inc.com, December 1, 2005, http://www.inc.com/resources/office/articles/20051201/morgan.html/.

Osono, Emi, Norihiko Shimizu, and Hirotaka Takeuchi, *Extreme Toyota—Radical Contradictions that Drive Success at the World's Best Manufacturer*, Wiley, Hoboken, NJ, 2008.

Rummler, Geary A., and Alan P. Brache, *Improving Performance: How to Manage the White Space in the Organization Chart*, Jossey Bass Business and Management Series, Jossey-Bass, San Francisco, 1995.

Sarkar, Debashis, "Customer Guarantees," *The Smart Manager*, August–September 2005, Vol. 4, no. 5.

Sarkar, Debashis, "Kaikaku—The Improvement Blitzkrieg," *The Smart Manager*, June–July 2006, Vol. 5, no. 4.

Sarkar, Debashis, *5S for Service Organizations and Offices—A Lean Look at Improvements*, ASQ Press, Milwaukee, WI, 2006.

Sarkar, Debashis, *Lean for Service Organizations and Offices—A Holistic Approach for Achieving Operational Excellence*, ASQ Press, Milwaukee, WI, 2008.

Schein, Edgar, "Coming to a New Awareness of Organizational Culture," *Sloan Management Review*, 1984, Vol. 25, no. 2, 3–16.

Shiba, Shoji, and David Walden, *Breakthrough Management*, CII, New Delhi, India, 2006.

Signet Research & Consulting, "After Action Reviews (AAR) and the Complete Action Review Cycle (ARC)," 2014. http://www.signetconsulting.com/methods_stories/proven_methods/after_action_reviews.php.

Smith, Gerald F., *Quality Problem Solving*, Prentice Hall India, New Delhi, India, 2000.

Sobek, Durward K., and Art Smalley, *Understanding A3 Thinking—A Critical Component of Toyota's PDCA Management System*, Productivity Press, Boca Raton, FL, 2008.

Swank, Cynthia Karen, "The Lean Service Machine," *Harvard Business Review*, October 2003, Vol. 81, no. 10, 123–129, 138.

Womack, James P., and Daniel T. Jones, *Lean Thinking—Banish Waste and Create Wealth in Your Corporation*, Free Press, New York, 1996.

Womack, James P., and Daniel T. Jones, *Lean Solutions—How Companies and Customers Can Create Value and Wealth Together*, Free Press, New York, 2005.

Index

Note: Page numbers ending in "f" refer to figures. Page numbers ending in "t" refer to tables.